D0687613

The City and the Coming Climate
Climate Change in the Places We Live

In the first decade of this century, for the first time in history, the majority of the planet's population resided in cities. We are an urban planet. If ongoing changes in climate are to have an impact on the human species, most of these impacts will play out in cities. This fact was brought into full relief in the summer of 2003, when more than 70,000 residents of Europe perished in one of the most prolonged and intense heat waves in human history. The final death toll would exceed that associated with any Western European or American conflict since World War II or any other natural disaster to have ever struck a region of the developed world, and the vast majority of these deaths occurred in cities. Studies in the aftermath of the heat wave would show that not only had climate change increased the likelihood of such an extreme event but also that the intensity of the heat had been greatly enhanced by the physical design of the cities themselves, exposing residents of cities to a much greater risk of illness or death than others.

This book is the first to explore the dramatic amplification of global warming under way in cities and the range of actions that can be taken to slow the pace of warming. A core thesis of the book is that the principal strategy adopted to mitigate climate change – the reduction of greenhouse gases – will not prove sufficient to measurably slow the rapid pace of warming in cities. The reason for this is that the primary driver of warming in cities is not the global greenhouse effect but rather the loss of trees and other vegetative cover to development and the emission of waste heat from industries, vehicles, and buildings. Rising levels of heat constitute the single greatest climate-related threat to human health, now accounting for more deaths per year than all other forms of extreme weather combined, and are contributing to the loss of life and unprecedented infrastructure disruptions in the present time period – not decades in the future.

Brian Stone explains the science of climate change in terms accessible to the nonscientist and with compelling anecdotes drawn from history and current events. The book is an ideal introduction to climate change at the urban scale for students, policy makers, and anyone who wishes to gain insight into an issue critical to the future of our cities and the people who live in them.

BRIAN STONE, JR., is an Associate Professor in the School of City and Regional Planning at the Georgia Institute of Technology, where he teaches in the area of urban environmental planning and design. His program of research is focused on climate change at the urban scale and is supported by the U.S. Centers for Disease Control and Prevention, the U.S. Environmental Protection Agency, and the U.S. Forest Service. Stone's work on urbanization and climate change has been featured on CNN and National Public Radio and in print media outlets such as *Forbes* and *USA Today*. Stone holds degrees in environmental management and planning from Duke University and the Georgia Institute of Technology.

The City and the Coming Climate

Climate Change in the Places We Live

BRIAN STONE, JR.
Georgia Institute of Technology

CAMBRIDGE
UNIVERSITY PRESS

CAMBRIDGE UNIVERSITY PRESS
Cambridge, New York, Melbourne, Madrid, Cape Town,
Singapore, São Paulo, Delhi, Mexico City

Cambridge University Press
32 Avenue of the Americas, New York, NY 10013-2473, USA

www.cambridge.org
Information on this title: www.cambridge.org/9781107602588

First published 2012

Printed in the United States of America

A catalog record for this publication is available from the British Library.

Library of Congress Cataloging in Publication data
Stone, Brian, Jr.
The city and the coming climate : climate change in the places we live /
Brian Stone, Jr.
    p.   cm.
Includes bibliographical references and index.
ISBN 978-1-107-01671-2 – ISBN 978-1-107-60258-8 (pbk.)
1. Climatic changes – Economic aspects.   2. Climate change mitigation.
3. Urbanization.   4. Cities and towns.   I. Title.
QC903.S86   2012
363.738′74–dc23   2011044842

ISBN 978-1-107-01671-2 Hardback
ISBN 978-1-107-60258-8 Paperback

To my father, Brian Stone, Sr. – Defender of the undefended, Mayor of Peachtree Street, and still-shining light to his family.

Contents

The color plate section follows page 86.

Acknowledgments

The origins of this book lie not in a large city but in a small one. I first learned that cities could influence their own climates in Durham, North Carolina, in a course taught by John Vandenberg in the Nicholas School of the Environment at Duke University. Since that time, more than 15 years ago, I have remained fascinated by the idea and am greatly indebted to John for his early and continuing support of this work. Over the years, many other teachers and colleagues have enabled and inspired this work. Thanks to Michael Rodgers at Georgia Tech and John Norman at the University of Wisconsin–Madison for their great patience in assisting a humanities major in tackling the complexities of urban climatology and for their direct collaboration on papers that have informed the book. Thanks to Michael Elliott at Georgia Tech for guiding this work in its earliest stages, helping to ground it in the realm of policy, and reviewing portions of the book.

I am greatly indebted to Matt Lloyd at Cambridge University Press for recognizing the value of this project from its inception and for serving as the book's editor. Thanks to Jeremy Hess and Daniel Rochberg, both at Emory University, for reviewing portions of the book and providing critical feedback. A debt of gratitude is also owed the book's several anonymous peer reviewers.

The book draws on a number of studies undertaken at the Urban Climate Lab at Georgia Tech and in collaboration with colleagues elsewhere. My profound thanks to Jason Vargo for his work in maintaining our database of urban climate trends, his expertise in graphic design, and his countless hours devoted to checking and rechecking the innumerable bits of data incorporated into the book. I am likewise appreciative of Dana Habeeb's assistance in compiling data and relevant studies for inclusion in the book. Thanks to Howard Frumkin at the University of Washington and Jeremy Hess for their valued collaboration on the extreme heat research highlighted in Chapter 3. Thanks to Claire Thompson for her assistance in gathering materials on the European heat wave for the book's prologue. I am furthermore indebted to the authors of the many studies cited throughout the book who have contributed to a vitally important and growing literature on climate change at the urban and regional scales.

Last, I am grateful for family and friends, who have supported and encouraged this sometimes all-consuming endeavor over the past several years. Above all, I am greatly appreciative of my wonderful wife, Elizabeth, for her willingness to provide whatever time and resources the project demanded; for her gifted editorial skills, often rendered in lieu of much-needed sleep; and for her endless patience in entertaining daily updates on the book's progress and impediments. I am likewise indebted to my sons for sharing their father with the unseen sibling of a book project that sometimes encroached on weekends and evenings. They have remained, above all else, the chief inspiration for the work.

Prologue: La Canicule

The unseasonably warm weather throughout Great Britain in the spring of 2003 was embraced by a population accustomed to the persistently overcast and damp conditions of a Northern European winter. Although still cold, February and March of that year had yielded an unusual number of sunny days, with relatively few rainstorms and periods of overcast skies. In April, Britons flocked to beach communities for the Easter holiday, taking advantage of temperatures reaching into the 70s, a generous 10 degrees above normal for that month.[1] As one media report noted at the time, "An unexpected glimpse of sunshine could brighten the weekend for millions of people in southern England" [1]. Another concluded that not even Miami, Florida, could muster the same tropical conditions experienced in Northern Europe at times that year. Mother Nature, it seemed, was smiling on the island kingdom.

The explanation for Britain's good fortune was to be found in the presence of a stationary high-pressure weather system centered over Scandinavia, which was drawing in warmer air from farther afield and elevating temperatures across Europe. The warm weather that Easter weekend was enjoyed in several European capitals, where long-shut windows were opened to blue skies and winter layers removed. In the spring of most years, momentary glimpses of the Sun over Northern Europe are to be celebrated; this year, however, the Sun was here to stay.

Although no one recognized it at the time, Europe was experiencing the early stages of a heat wave so extreme that it would far surpass any comparable weather event in more than three centuries of record keeping. Since temperature observations were first maintained in 1659, a period when Louis XIV ruled France and the Pilgrims occupied

---

[1] Unless otherwise indicated, all temperatures are reported in degrees Fahrenheit.

Plymouth, not a single summer had produced temperatures so intense and over such an extended period of time. Maintained by a succession of stationary weather systems over the Northern Atlantic and Central Europe, conditions of excessive heat and drought would persist for almost eight months. By summer's end, the heat wave had reduced ancient rivers to non-navigable streams, consumed in fire an area larger than some European nations, and claimed more lives than the United States lost in a decade of warfare in Vietnam.

The heat wave of 2003 would constitute the single most catastrophic weather event to be visited on Europe – and, arguably, any modern nation – during the period in which weather observations have been recorded. Less than a decade since its occurrence, however, many reading these pages will not recall having heard of the event.

This perhaps was the central lesson of the crisis: heat kills quietly.

By April 2003, the heat wave was well underway. Long accustomed to soggy winters and persistent overcast conditions, Western and Central Europe had instead received relatively little precipitation during the month of February. For example, rainfall that month in Frieburg, Germany, was about 60% less than the long-term average, and would continue to fall significantly below average levels for the following seven months. By May, large expanses of Europe were experiencing a rainfall deficit between 50–75% of the long-term average, symptomatic of a deepening drought.

Although heat waves are most directly characterized by a succession of excessively hot days, such elevated temperatures over an extended period are often a response to longer term deficits in rainfall. Moisture, in this sense, is a critical regulator of climate. As the landscape is depleted of moisture, a larger percentage of energy received from the Sun is converted into heat at the Earth's surface, serving to increase temperatures. By working against the upwelling of water vapor to the atmosphere, the subsiding air mass of a high-pressure weather system over Europe was effectively robbing the atmosphere of moisture needed for rainfall. As can be experienced in any desert climate, a landscape starved of moisture is prone to temperature extremes.

By May, the anomalous warm spells of March and April had developed into a persistent warming trend, with temperatures across much of Europe registering more than 7°F higher than long-term averages for the month. By June, the development of a fully fledged heat wave was widely apparent. Temperatures soared in the south of France, where daily maximum temperatures were averaging more than 10°F above normal and reached as high as 104°F. Switzerland suffered through its hottest June in more than 250 years of record keeping, with nighttime temperatures in Geneva rarely falling below 80°F. As temperatures climbed to 90°F in Britain, local bookies starting taking bets on a statistical outcome that had never before seemed worthy of a wager: that England could reach an unimaginable 100°F. By July, the odds of such an event were 14–1, and falling [2].

The first signs of the developing public health emergency were the gradually increasing number of patients arriving at large urban hospitals complaining of general fatigue and shortness of breath. Because such symptoms are indicative of a wide range of potential ailments, they were hardly sufficient to alert doctors and public health officials to the wave of fatalities that was soon to come. As a French official observed somewhat bitterly at the height of the heat wave, "People don't come in with 'dying of heat' on their foreheads" [3]. Yet those symptoms would later be linked directly to the excessive temperatures outdoors and in buildings lacking air conditioning. And the most vulnerable populations were already succumbing to the heat. In Paris, a social services organization would report the deaths of two homeless men from no obvious cause following a string of excessively hot days in early August. The cause would soon become apparent.

The human body can tolerate surprisingly high temperatures. With a warm-blooded core temperature of 98.6°F, the very baseline temperature for human survival is considerably warmer than most outdoor environments. However, the tolerable deviation from this core temperature for more than a brief period of time is quite limited, and such deviation is managed more easily in environments that are somewhat cooler than this threshold. As environmental temperatures approach 98.6°F, the human body is increasingly required to cool itself through the release of water in the form of perspiration. At temperatures above this threshold, sweating assumes the urgency of life support.

The importance of moisture in cooling the human body is really no different than its role in regulating the temperature of ambient air. Moisture offsets heat gain through the process of evaporation. As water evaporates, it makes use of heat energy received from the Sun or ambient air to convert water to water vapor. Through this conversion process, heat energy that would otherwise contribute to an increase in skin temperature is effectively locked up in the vaporized water molecules and transported away. Moisture availability in the human body, therefore, is a critical factor in the body's ability to cope with rising environmental temperatures.

High temperatures stress the human body by overworking the heart and other organs. As temperatures rise, blood must be pumped faster throughout the body to distribute the water needed for perspiration. As long as these lost fluids are continuously replenished, the body can continue to cool itself through perspiration indefinitely. However, the heart is increasingly stressed by the need to circulate blood at an elevated rate. Other organs can be stressed by the loss of sodium, which is also released through the sweating process. If fluids are not replenished, the circulating blood begins to thicken, elevating the risk of clotting and, in the most severe cases, heart attack or stroke. If salts are not replenished, muscles begin to cramp and organs cease to function normally.

The significance of moisture availability to thermoregulation of the human body illuminates a central paradox of heat waves: many victims succumb to heat exhaustion or heat stroke during the night, when temperatures are lower than during the day. Because heat waves are characterized not only by above-average maximum temperatures during the day but also by above-average minimum temperatures during the night, the body may not be afforded sufficient time to recover from heat exposure during a 24-hour period. This is particularly true for vulnerable populations who lack access to air conditioning during the nighttime hours. If the body continues to perspire during the night as a person sleeps, the heart must continue to function at elevated levels and fluids must continue to be replenished. A failure to do so, particularly over a string of several days during a heat wave, can result in a gradual increase in the body's core temperature. If this temperature increases beyond a very limited range – 4 to 5 degrees – the likely result is death.

If a gradual increase in the number of patients admitted to hospitals for heat-related ailments was not sufficient to alert public health officials to an imminent emergency, stress on the natural and physical environments from the relentless heat should have left few doubts as to the severity of the heat wave. By the middle of July, the intensity of the drought, now in its sixth month, was revealing itself in a growing number of wildfires and falling water tables. Sparked by lightning and, in several instances, arson, fires in France, Portugal, Switzerland, and Spain spread rapidly through wilting croplands and forests. By the end of July, 2,300 firefighters were combating 72 separate blazes in Portugal alone [4]. Fires raging across the French Riviera were characterized as the worst in a generation.

In response to the worsening drought, some of Europe's major rivers fell to historically low levels, greatly limiting the availability of water for irrigation and shipping. By July, the River Po, which provides irrigation water for about a third of all agriculture in Italy, was measured to be 24 feet below its normal flow, its lowest level in 100 years [5, 6]. Diminished levels of the Danube required ships departing Austria to be loaded at no more than 50% capacity [7], and shipping elsewhere on the river was forced to cease altogether [6]. Water flows became so diminished along several Belgian rivers that not even kayakers could safely navigate the channels. Elsewhere, receding waters would reveal unanticipated hazards for shipping and recreation. Four unexploded bombs from World War II were exposed by the retreating waters of Lake Constance in Germany, and the long entombed hulls of ships sunk by bombs during the same conflict jutted forth from the Danube near the Balkans.

Much like the human body, urban infrastructure can become stressed during prolonged periods of intense heat exposure. It is the failure of infrastructure in the form of transportation and electrical power systems that can pose one of the most significant – and the least anticipated – threats of rising temperatures to urban populations. One unanticipated outcome of the 2003 heat wave was the buckling of rail systems and roadways. In July, a freight train carrying a shipment of Guinness beer outside of London derailed as a result of a section of steel rail that had been physically warped by the intense heat. Because all metals are subject to expansion or contraction in response to temperature changes, the steel used for train tracks must be engineered to tolerate a wide range of thermal conditions. An article from the BBC shortly after the derailment explains well the hazard of heat for rail transit systems:

> To prevent [derailments]. . . . the track is "pre-stressed" or stretched, so that when it gets hot, the metal cannot expand any further. It is designed to cope with temperatures up to 30 degrees centigrade (86°F). "Ah," say the armchair engineers, "why not stretch it more so it can take more heat?" The problem is that the stretching makes it more likely to crack in cold weather – British track can take our coldest winters. The main problem is that it is currently much hotter than normal, not just in Britain but across the rest of Europe. The temperature of the track can reach 50 degrees (122°F) [8].

Similar to steel rails, the concrete or asphalt of paved surfaces may also expand beyond engineering thresholds in response to intense and prolonged heat. Over a threshold temperature, asphalt can literally melt, while concrete ruptures, rendering streets impassable. The buckling of rail and roads in several instances across Europe in the summer of 2003 would intermittently delay and force closures of critical infrastructure. In some instances, the heat alone was sufficient to shut down transit systems absent a failure in infrastructure or equipment. Suffering through cabin temperatures as high as 120°F, a tram operator in Helsinki, Finland, lost consciousness, causing him to run a red light and strike a motorcyclist (killing him) before crashing into the wall of a department store [9]. Although no one on the tram was critically injured, the incident revealed yet another unanticipated threat of extreme heat to urban populations: transport system failures.

Undoubtedly the most critical infrastructure during a heat wave is that involved in power generation and transmission. Air-conditioned environments must be accessible to enable the human body to cool during periods of intense heat. Yet, the very conditions that give rise to an urgent demand for electricity can greatly stress the infrastructure delivering that electricity. This was certainly the case in July and August 2003, when the demand for energy soared throughout Europe to meet the cooling needs of homes, offices, and hospitals.

In some instances, operating temperatures within nuclear power plants approached thresholds that would have required an immediate shutdown. One plant near Strasbourg, France, had to be manually hosed down to avoid the violation of safe operating standards as plant temperatures reached 119°F [10]. Elsewhere, plants were taken offline entirely because of an insufficiency of cooling water from drought-stricken rivers. During the height of the heat wave, France, Europe's chief exporter of energy, was required to reduce energy exports by more than 50% to account for the shutdown of numerous power plants and the spiking domestic demand for electricity [11]. Together, these events highlight an obvious but widely underappreciated threat of extreme weather to human health: the reliability of essential infrastructure tends to decline at the very moment population vulnerability rises.

By August, the European heat wave had reached the height of its intensity. During the first two weeks of that month were recorded temperature extremes that had not been experienced in almost 350 years of record keeping – surpassing the longest running set of direct temperature measurements in existence. On August 10, the United Kingdom would reach an all-time record temperature of 101.3°F, exceeding by several degrees the previous high set a few decades earlier. Michael McCarthy, a columnist for *The Independent*, described well the significance of this event:

> The point about 100 was, it was off the map. . . . It was off the scale completely. But never mind it not being in the meteorological history; it wasn't in the cultural history. There was no cultural reference point for it: no stories, no memories, no jokes, no newspaper headlines. In the temperate Britain for which we all have an instinctive feel, this land of showers and cool summers, this land on the latitude of Labrador only kept from freezing by the Gulf Stream, an air temperature of 100° Fahrenheit represented an unknown country, an elsewhere. The round figure helped with that, to be sure. In Centigrade terms, 100°F is 37.8, and of course 37.8 as such isn't any sort of figure the mind will register, any sort of boldly-marked frontier whose breaching will seem significant; but once you represent it in Fahrenheit, the move up from two digits to three has a symbolical significance of real power. It is a true cultural border, as well as a meteorological one. Above 100 is new territory [12].

Britain was not the only country to reach new territory. On August 11, the high temperature in Switzerland, land of the glacial Alps, would reach an unimaginable 107°F. Between August 7 and 14, minimum temperatures in Paris never dipped below the mid-70s, with the city's hottest night on record suffered through on August 11–12, registering a minimum temperature of almost 80°F. Satellite measurements of temperature across Europe during the height of the heat wave reveal the extremity of the event, with many areas exhibiting temperatures almost 20°F above normal (Figure P.1).

To fully appreciate the gravity of these numbers, it must be noted that air conditioning remains a luxury throughout much of Europe – particularly in individual homes. In 2004, for example, only about 7,500 homes in the United Kingdom – far below the size of a single London borough – were found to be equipped with central air conditioning [13]. Likewise, air conditioning in France, even in the south of France, is not commonly found in homes. As a result, during a two-week period in August, many Europeans were enduring outdoor temperatures in excess of 100°F during the day and close to 80°F in the dead of night, with unventilated indoor temperatures reaching much higher. Because the human body generally can only endure about 48 hours of exposure to excessive heat before suffering effects of heat exhaustion or stroke, the heat wave

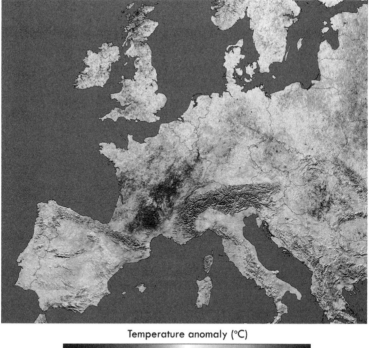

Temperature anomaly (°C)

-10   -5   0   5   10

Figure P.1   Temperature anomalies (degrees above or below normal)
across Europe on July 31, 2003. *Source:* Adapted from the NASA Earth
Observatory http://earthobservatory.nasa.gov/NaturalHazards/view.php?
id=11972. (*A color version of this figure can be found in the color plate section
after page 86.*)

of August 2003 had become for many a deadly serious weather event. It
had become, in short, a natural disaster.

As with all natural disasters, the number of lives lost in a heat wave
has as much to do with the adequacy of the emergency response system
in place as with the severity of the event itself. In August 2003, European governments were poorly prepared for the public health emergency
brought on by the heat wave. Much of the insufficiency of the response
was a product of the governments having never contemplated heat as
a genuine threat to their populations and therefore having not developed emergency response plans. Also problematic was the timing of the
event, which occurred during the height of the summer holiday season

in Europe, when much of the workforce, including medical personnel, was on vacation.

The rapidity with which a public health emergency was unfolding became apparent in the first days of August. At a number of hospitals in France, patient admissions spiked by 20% to 30% over a two-day period, quickly overwhelming understaffed emergency personnel. In one hospital, 20 patients died from heat exhaustion or stroke in a single weekend – a 10-fold increase in the typical number of deaths. As the initial death toll attributed to the heat wave in France reached 50, the president of the French Association of Accident and Emergency Doctors sounded an alarm in the media. "The weakest are dropping like flies," he protested. Government officials "dare to say these deaths are natural. I absolutely do not agree. No statistics are being gathered. There is no general information, nothing" [3].

At the time, it was observed that the majority of those dying were over the age of 65, and a disproportionate number of these individuals lived alone. Less perceptive to the initial symptoms of heat-related health conditions and less mobile in general, older populations have been found in past heat-wave events to be more susceptible to heat exhaustion and stroke. This baseline vulnerability was aggravated in the summer of 2003 because many family members were away on holiday during August and were unable to check in on elderly relatives during the height of the event. It was also found that elderly victims were less likely to open windows at night – a fact attributed to a greater fear of crime – and thus were exposed to dangerously high indoor temperatures.

By the second week of August, the swelling number of patients suffering from heat-related illnesses was overtaxing medical facilities throughout European cities. Yet, unconvinced of the severity of the event, government officials refused to recall vacationing doctors and medical staff home to their hospitals. The inability of the French medical system to cope with the increased numbers of patients is illustrated by the experience of a woman who had taken an elderly aunt suffering from a heat-related pulmonary condition to a hospital. She recounts the following:

> At first [my aunt] was put in an air-conditioned revival room but then she was abruptly transferred to a ward where it was 50°C [122°F]. I talked to two nurses. One said: "I don't have time to bother with her." The other said: "Get her out of here." But the doctors would not let her go. Three days later, she died [14].

What such anecdotal observations cannot convey is the sheer number of people who were dying from heat exposure during the middle of August. Studies commissioned by European governments after the heat wave would find that literally thousands of victims had died over just a handful of days as extreme temperatures persisted unabated. In France alone, it is estimated that approximately 2,200 heat-related deaths occurred on a single day in August [15]. Another 600 had perished that day in Spain and more than 250 in just four Italian cities [16, 17].

The rate at which people were succumbing to the heat in hospitals and in their homes was overwhelming the governments' ability to manage the most tragic aspect of a natural disaster: how to dispose of the growing number of corpses. A newspaper account from late August describes the inability of morgues and undertakers to cope with the bodies of heat-wave victims:

> Nine refrigerator trucks, containing more than 100 unclaimed bodies, have been parked under police guard in a municipal car park in Ivry-Sur-Seine, a southern suburb of the capital, it was revealed yesterday. Others are being kept in a cold store at the Rungis vegetable market and in a morgue normally used for murder victims. Forty other bodies have already been buried in individual graves in the paupers' section of a cemetery east of Paris. City officials said that the bodies could be exhumed and buried elsewhere, or cremated, when relatives are traced. They denied press reports that some bodies had been placed in a mass grave [18].

Yet those press reports were not far off the mark. Under city law, any body unclaimed by relatives after a period of six days must be placed in a "common section" of a municipal cemetery in Paris. Despite an extension of this period in the aftermath of the heat wave to 10 days, many victims of the heat wave, never identified or claimed by family members or friends, were interred in these unmarked graves (Figure P.2).

Unadorned by headstones or statuary, these already forgotten paupers' graves in the far reaches of Parisian cemeteries constitute the first undeniable monuments to a shifting environmental order, and attest to a simple but largely unappreciated truth: our affluence, science, and civil institutions are no match for changing weather. So extreme were the events of August 2003 in France, as throughout Europe, that the language used to describe the event was altered in simple but profound ways. In much the same way as the date 9/11 has come to assume a deeper meaning to Americans, the French term for heat wave – "canicule" – would come in the aftermath to refer to a particular,

Figure P.2   Unmarked graves for unidentified heat victims in Paris.
Source: Franck Prevel, Associated Press, 2003.

historically significant episode, as opposed to any stretch of unusually hot days. No longer were the events of that summer referred to in the media and popular language as "une canicule" – this particular summer would be thereafter referred to simply as "La Canicule."

In the months following the heat wave, postevent assessments would document massive social, economic, and environmental impacts. By the end of the summer, more than 25,000 fires had consumed a total of 647,069 hectares across Portugal, Spain, France, Italy, Austria, Finland, Denmark, and Ireland – an area roughly equivalent to that of Luxembourg [11]. Portugal alone would lose an estimated 10% of its total forestland [19]. Agricultural losses were unprecedented. The excessive heat and drought reduced fodder harvests in France by an estimated 60% and by at least 30% in neighboring countries [20]. One of the most productive agricultural regions in Italy reported a 40% to 50% drop in olive production and a 40% to 100% reduction of the peach, apricot, and grape yields [5], with the costs of produce on the shelf in Britain rising by 40% [21]. The limited evidence available suggests that the impact of the heat on livestock was substantial: it was reported that 80,000 chickens perished on a single farm in England [22]. Overall, economic losses from the heat wave across

Europe were estimated to be about $13 billion [20], a figure that is certain to underestimate the true costs.

The most consequential statistics, however, concerned the loss of human life. European governments were astounded to discover the true number of their citizens who had perished from the heat over the course of several weeks in a single summer. A joint study commissioned by the European Union (EU) would show through comparisons of fatality rates in June through September 2003 with previous months, or with the same months in previous years, that tens of thousands of excess fatalities had resulted from the extreme heat [23].

The epicenter of the disaster was in the countries of France and Italy. Initially estimated in the immediate aftermath of the heat wave to have suffered 5,000 fatalities, French officials would later discover this number to underestimate the actual death toll by about 300%. During the period spanning June through September 2003, France suffered a staggering 19,490 excess deaths from the heat wave – almost 12 times greater than the number of deaths typically experienced during these months. Italy was found to closely match this grim total, with 20,089 citizens having died from the heat. What is most remarkable about these numbers – and perhaps most foreboding – is that almost 40,000 citizens had died from hot weather in two of the most affluent and medically advanced societies in the world.

In all, the EU estimated that more than 70,000 citizens of 12 countries died from heat-induced illnesses over a four-month period in the summer of 2003. This number represents more fatalities than have resulted from any EU or American conflict since World War II or any natural disaster (e.g., hurricanes, earthquakes, and floods) to have ever struck a developed nation. It dwarfs the 1,800 deaths attributed to Hurricane Katrina in 2005 and effectively renders trivial the 900 lives lost during the highly publicized SARS epidemic that struck in the same year as the heat wave, an event giving rise at the time to virtual hysteria. Americans would need to experience more than 20 terrorist attacks equivalent in destruction to 9/11 before such a death toll would be approached. Yet the global response to this climate event, an event that reveals more about the profoundly changing environment in which we now live than any other yet endured, has largely been one of indifference. Although numerous books have been published and movies produced on recent disasters such as Katrina and the SARS outbreak – at the time of this writing, the U.S. Library of Congress held more than 200 books on Hurricane Katrina alone – to date, not a single book has been published on the 2003 heat wave.

Not one.

This is a book about climate change in cities. Although my explicit focus is not on the catastrophic European heat wave of 2003, the events of that summer highlight a central truth about climate change that is often lost in the global- and future-oriented debate over a warming environment: the impacts of climate change at the urban scale are profoundly greater than the impacts of climate change at the global scale. And a second truth: the impacts are here with us today.

One of the key findings of postevent assessments following the 2003 disaster was that the vast majority of those who perished in the heat wave lived in cities. This fact is not explained by the larger percentage of the national populations residing in cities across Europe because the rate of heat-wave fatalities among residents of large urban areas such as Paris and London was greater than that among residents of smaller towns or rural areas. And this fact is at odds with the general presumption of superior health care in cities, which provide far greater access to emergency medical facilities. The heat wave claimed a disproportionate number of lives in cities simply because the cities were hotter than rural areas – substantially hotter.

Cities do not cause heat waves – they amplify them. Because of the greater prevalence of mineral-based building materials, such as stone, slate, concrete, and asphalt, cities absorb and retain substantially more heat than rural areas characterized by more vegetative cover. Known generally as the "urban heat island effect," this phenomenon keeps cities warmer by several degrees than surrounding countryside throughout the year. However, during unusually hot days, the divergence between urban and rural temperatures can be much greater, literally tipping the balance between an unpleasantly hot day in one environment and a public health emergency in another.

This observation is illustrated well by data obtained during the 2003 heat wave. As part of an ongoing research study unrelated to the heat wave itself, an extensive network of meteorological instruments was in place in and around Strasbourg, France, throughout the heat-wave event [24]. Data from this network showed that the increase in heat-index values at the height of the event, a measure that accounts for both temperature and humidity and most closely captures the physio-logical impacts of heat on the human body, was about 30% greater in the downtown district than in the surrounding countryside. Nighttime heat-index values, which provide the most direct indicator of the body's ability to cool down during a heat wave, were at times measured to be 50% greater in the urban center. What these measurements show is that the effects of the heat wave for urban residents were as much as 50% greater than for rural residents just a few miles away. The significance of these differences is hard to overemphasize: one's decision to remain in a city during a heat wave can quite literally mark the difference between life and death.

In the first decade of this century, for the first time in history, the majority of the planet's humans resided in cities. We are an urban planet. If ongoing changes in climate are to have an impact on the human species, most of these impacts will play out in urban environments. Yet climate science to date has provided very few insights into how cities, in particular, will be influenced by climate change. Continuously framed as a global phenomenon, with implications for the planet as a whole, the climate change of peer-reviewed scientific papers and international accords does not seem to be taking place anywhere people actually live. In fact, the moment a climate-related event with tangible geography occurs, such as a heat wave or hurricane of unprecedented intensity, we are quickly told that no single event can be proven to be an indication of climate change. For the nonscientist attempting to formulate an opinion on the issue, the likely outcome is not surprising: if climate change is not happening in the places we live, it's not happening.

More problematic for cities than the framing of the issue, however, is the uniform adoption of the global scale as the legitimate basis for scientific inquiry. Preoccupied with measuring the rate of temperature change at the scale of the planet as a whole, we have largely overlooked the rate at which climate is changing in cities. Indeed, as examined in the following pages, temperature data from urban weather stations are statistically adjusted in the global temperature datasets employed by climate scientists to measure global warming. Were these temperature measurements not modified, we would find that the environments in which we live are actually warming at a substantially higher rate than the planet as a whole, with troubling implications for anyone who lives, works, or owns property in cities.

Above all, it is the rate at which climate is changing in cities that most clearly illuminates the lessons of the European heat wave for urban governments and residents. Were such an event to remain a statistical improbability, cities could be forgiven for prioritizing other critical needs above preparations for combating climate change. Studies focused on this question following the heat wave would show that, absent the influence of human-induced warming, an event approaching the intensity and duration of the 2003 heat wave would indeed remain quite rare, occurring, on average, once every thousand years. Yet, assuming global temperatures continue to rise at the rate of recent decades, the frequency of such a heat wave increases substantially – so much so that, by the year

2040, such heat waves may be expected to occur every year [25, 26]. Such a world seems hardly imaginable: temperatures of sufficient intensity and duration to physically warp the steel of railroad tracks and melt the asphalt of streets – every year.

For those of us who give ourselves better than even odds of being here in 30 years, a principal lesson of La Canicule is undeniable: this is not our grandchildren's problem alone.

1

Keeling's Curve

One of the earliest descriptions of the global greenhouse effect to appear in the popular media can be found in a 1953 edition of *Popular Mechanics*. The brief two-paragraph article published more than a half-century ago describes with surprising precision the fundamental workings of a scientific phenomenon that would soon become one of the most intensely studied in human history. The piece explains, "Earth's ground temperature is rising by about $1\frac{1}{2}$ degrees a century as a result of carbon dioxide discharged from the burning of about 2,000,000,000 tons of coal and oil yearly.... This discharge augments a blanket of gas around the world which is raising the temperature in the same manner glass heats a greenhouse"[1]. The article goes on to predict a doubling of atmospheric carbon dioxide levels and a rise in global temperatures of about 4% by 2080, which is not far off the mark from today's best estimates.

That an article from the 1950s could describe with measurable accuracy the workings of a phenomenon that would not be formally acknowledged by the U.S. Academy of Sciences for another several decades is remarkable. Yet perhaps the article's central prescience is suggested by its placement within the magazine, appearing on page 119 of the August edition and trailing a piece titled, "Dutch Entertainer Rides Tiny Bike." Viewed by the magazine's editors as less newsworthy than a shopworn circus act, the global greenhouse effect and its implications for human life are viewed today by many in a similar light. But this outcome results not from any degree of scientific uncertainty.

The 1950s marked an important era in the realm of climate research due not to the recognized association between the greenhouse effect and fossil-fuel emissions – that linkage had been established decades earlier – but rather to the installation of a scientific device on a volcanic mountaintop in Hawaii. There, at the Mauna Loa Observatory established by the U.S. Weather Bureau, Charles Keeling would install

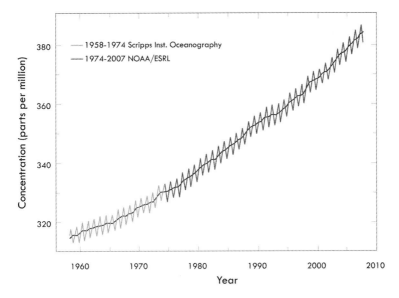

Figure 1.1   Measurements of atmospheric carbon dioxide at the Mauna Loa Observatory. *Source:* NOAA http://www.esrl.noaa.gov/news/2007/ 50YearCO2Record.html.

and monitor one of the first manometers, a device designed to measure atmospheric carbon dioxide (CO_2) and developed for long-term atmospheric research.

A research scientist at the Scripps Institution of Oceanography in California, Keeling was leading an effort to determine whether the concentration of carbon dioxide in the atmosphere was uniform across the Earth. In concert with observations from other weather stations, the Mauna Loa manometer would soon confirm the relative uniformity of CO_2 levels around the planet. The significance of this discovery was its demonstration that carbon dioxide is a globally diffuse gas, such that higher emissions in one region of the planet would ultimately increase atmospheric concentrations in all regions of the planet. Yet, over time, these measurements would reveal an unanticipated and more important observation about the concentration of CO_2 in the atmosphere: not only was CO_2 globally diffuse, its levels also were rising steadily, year after year, and rising fast.

Keeling's measurements, tracing an annually fluctuating but gently increasing curve every year since 1958, provide perhaps the most visually compelling window into the role of human activity in altering the Earth's climate (Figure 1.1). Yet, Keeling's curve would not mark the discovery of the global greenhouse effect. Rarely understood outside of scientific or climate-policy circles, and arguably a key impediment to further progress on the issue today, is that the physical mechanism of the global greenhouse effect has been settled science for more than a century. Although the degree to which human activities were enhancing the greenhouse effect would remain subject to credible debate until the end of the 20th century, the role of greenhouse gases in warming the Earth's atmosphere had been incorporated into basic texts on climatology many years before Keeling became interested in measuring the global distribution of CO_2.

The first theoretical description of this relationship was published in 1824 by a Frenchman named Jean-Baptiste-Joseph Fourier. A pioneer in the development of theory on heat diffusion, Fourier became interested in the mechanisms through which heat from the Sun is absorbed and maintained in the atmosphere. Fourier reasoned that heat is retained in the Earth's atmosphere, as opposed to simply radiating back into space, because of its absorption by specific atmospheric gases. The brilliance of Fourier's hypothesis was in his recognition that many atmospheric gases are largely transparent to radiation emitted by the Sun but opaque to radiation emitted by the Earth. As a result, incoming and reflected solar radiation is permitted to pass freely through the atmosphere to the Earth's surface, but outgoing terrestrial radiation, which is emitted at a different wavelength from that of sunlight, is trapped by the atmosphere. By trapping a portion of this outgoing terrestrial radiation, greenhouse gases in the atmosphere act as an insulating blanket around the planet or, in Fourier's terms, a "bell jar," thereby raising the Earth's temperature [2]. The physical principles described by Fourier are depicted in Figure 1.2.

Although he did not characterize it as such at the time, Fourier was describing physical properties that distinguish the atmosphere of the Earth from that of the Moon. Endowed with a dense atmosphere composed of numerous gases and reaching several miles into the sky, the Earth exhibits many physical properties that the Moon, encircled by a very thin, almost negligible atmosphere, does not. For example, one of these properties is a blue sky. When sunlight reaches the Earth's atmosphere, blue wavelengths are scattered by gases present in the atmosphere, giving the sky a diffuse blue color. As revealed by photographs from the Moon's surface, the absence of a dense atmosphere there produces, by contrast, a stark black horizon.

Just as the Earth's atmosphere interacts with incoming solar radiation, it traps outgoing terrestrial radiation, increasing surface

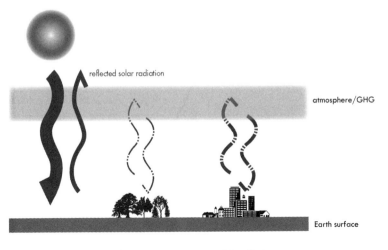

reflected solar radiation

atmosphere/GHG

Earth surface

▬▬▬ solar radiation   ▬■␣␣■▬ terrestrial radiation

Figure 1.2   The global greenhouse effect. GHG denotes greenhouse gases.
Sources: Subikto Tree font from Sub Communications Fonts and Little City
2000 font from Heather Daniel.

temperatures by about 60 degrees and rendering the planet suitable for
life. Effectively lacking an atmosphere, the Moon enjoys no comparable
insulating effect and is much colder, with average lunar temperatures
hovering close to zero.

If validated, Fourier's bell-jar hypothesis would hold the potential
to address many long-standing and substantive gaps in the developing field of atmospheric physics in the early 19th century. Yet valida-
tion would require direct observations of the physical mechanisms that
Fourier was describing, and the necessary instrumentation to make such
observations had not yet been invented. Most critically, science had yet
to produce a means of measuring the degree to which atmospheric gases
absorb or transmit different types of radiation. If gases such as water
vapor and carbon dioxide were capable of absorbing outgoing terrestrial
radiation, such an interaction should be demonstrable in a laboratory. In
the absence of such measurements, Fourier's hypothesis would remain
one of several competing theories of heat diffusion in the atmosphere.

The first direct measurements of the absorptive capacities of atmospheric gases would be provided by John Tyndall, an experimental physicist in the Royal Institution of Great Britain, in 1859. Having constructed
the world's first ratio spectrophotometer for the very purpose of testing
Fourier's hypothesis, Tyndall was able to measure the degree to which

specific gases known to be present in the atmosphere, such as oxygen, nitrogen, carbon dioxide, and water vapor, absorb or transmit radiation. Tyndall's measurements would show that although the atmosphere's most common gases – oxygen and nitrogen – were transparent to both solar and terrestrial radiation, less common gases, such as water vapor and CO_2, selectively interacted with different forms of radiation. Just as Fourier had surmised, these gases were transparent to radiation emitted by the Sun but opaque to radiation emitted by the Earth. The greenhouse effect, not to be so named for the better part of another century, was real.

In the same year that Tyndall was measuring the radiative properties of atmospheric gases in London, a seemingly unrelated development across the Atlantic, in Titusville, Pennsylvania, would soon heighten the significance of his work. Titusville had become famous for a large number of natural oil seeps found along a small stream known as Oil Creek. First used by Native Americans as weatherproofing for wigwams and in medicines, the oil rising to the surface had grown increasingly valuable by the 1850s, after the invention of the kerosene lamp earlier that decade and in response to globally declining supplies of whale oil. In 1859, Edwin Drake had been sent to Oil Creek by the Seneca Oil Company, based in New Haven, to develop a means of extracting the oil from the ground at a higher rate than the natural seepage. To do so, Drake would construct a primitive drilling derrick powered by a steam engine to drive experimental drill bits into the uncooperative Pennsylvania bedrock. After weeks of work, Drake's crew would finally strike oil at a depth of only 69 feet. Within a few years of Drake's success, the horizon of Titusville would be littered with hundreds of such primitive derricks, the world's first oil boom fully underway.

The industrialization of oil production in 1859 would open a second major front in the global campaign to extract and burn fossil sources of energy. Over the preceding century, coal had been mined at an ever increasing pace to fuel steam-generated power for the industrial revolution. By the time Drake struck oil in Pennsylvania, the United States alone was mining more than 15 million tons of coal a year, or about a half-ton for every man, woman, and child in the country. Widely used not only to power industry but also to heat homes, coal was a persistent presence in urban areas, where uncontrolled emissions of coal smoke blackened the afternoon sky, requiring street lamps to remain burning throughout the day and causing a range of respiratory ailments. Yet the longer-term implications of burning coal, now increasingly accompanied by the burning of oil, would pose in the mind of one 19th-century scientist a question that had previously seemed unimaginable: Could an increase in atmospheric carbon dioxide associated with fossil-fuel burning warm the atmosphere?

Shortly after Tyndall's death in 1893, a Swedish physicist named Svante August Arrhenius began to focus on the potential for changes in atmospheric CO_2 levels to bring about measurable changes in climate – a question implied by Tyndall's work but never explored by the scientist himself. Although not a climate scientist by training – he would later win a Nobel Prize for his work related to electrochemistry – Arrhenius sought out of a passing interest to calculate the influence of a doubling or a halving of atmospheric carbon dioxide on climate, describing the work to a friend as a "trifling" matter [2]. Over the period of a full year, Arrhenius carried out a series of laborious computations by hand to estimate the relative significance of CO_2 levels to temperature on the Earth's surface. In 1896, the physicist reported his results in a paper, now perhaps the most widely read of any he would publish, titled, "On the influence of carbonic acid in the air upon the temperature of the ground" [3].

Arrhenius's computations suggested that significant changes in atmospheric CO_2 could lead to profound shifts in global climate. By his estimate, a reduction of carbon dioxide levels by half could lower mean global temperatures by between 7 and 9°F, sufficient to plunge the globe into a new ice age. Conversely, a doubling of CO_2 could increase mean global temperatures by between 9 and 11°F, causing glaciers to retreat and a dramatic warming of the northern latitudes. Acutely aware of the rapid industrialization occurring across Europe and the growing emissions of CO_2 from coal burning, Arrhenius believed a warming trend was the more likely of the two scenarios and estimated that, at then-current rates of fossil-fuel combustion, a full 3,000 years would pass before a doubling of CO_2 would be possible [2].

Arrhenius speculated that a mild pace of warming might be desirable. As described by Christianson (1999),

> [T]he more he pondered the idea of atmospheric warming, the more it appealed to him. The projected increase in temperature, he informed those who attended his lecture, will "allow all our descendants, even if they only be those of a distant future, to live under a warmer sky and in a less harsh environment than we were granted." [2]

It is hard to overemphasize the precision of Arrhenius's computations almost a full century before a panel of global scientists would be convened to study the issue. Equipped with only the most rudimentary understanding of the many complexities involved in global climate, ranging from cyclical variations in solar output to the uptake of CO_2 by the world's oceans, and lacking entirely the modeling capabilities of computers to estimate the sensitivity of climate to changes in the atmosphere, Arrhenius had derived with pencil and paper a climate change projection that comes surprisingly close to today's best estimates. In a single academic paper, he had demonstrated mathematically the potential for human activity to alter the weather, proposed a new theory to explain the advance and retreat of the ice ages, and provided a metaphor that would be universally adopted to explain the phenomenon – comparing the effect of carbon dioxide on temperature in the atmosphere to that of the glass in a "hothouse."

Yet, with respect to one critical issue – what has become the definitive issue for the adaptive capacity of the global ecosystem – Arrhenius's computations were dead wrong. Unable from his late-19th-century vantage point to recognize the potential for Drake's drilling works in Titusville, coupled with a burgeoning industrial society, to facilitate the transference of billions of tons of ancient carbon from the Earth to the atmosphere over a period of decades rather than millennia, Arrhenius underestimated by a factor of 30 the likely time required for a doubling of CO_2 in the atmosphere. It is precisely in light of this miscalculation that Keeling's measurements from the slopes of Mauna Loa assume their unalterable significance. At the rate at which excess carbon was accumulating in the latter 20th century, global CO_2 levels would experience a doubling from preindustrial levels not in 3,000 years as Arrhenius had calculated but rather in a single century. The distant future envisioned with optimism by the prophetic scientist was already at his doorstep.

One of the key challenges presented by rising levels of carbon dioxide in the atmosphere is the relatively long period required for carbon to cycle out of the atmosphere and back into the ocean or terrestrial reservoirs, such as forests. As with other common elements on Earth, carbon is continuously recycled among the land, atmosphere, and oceans through a natural process; however, human activities are altering the rate at which carbon cycles through this process. A typical cycling of a carbon atom might follow this path. First, a molecule of carbon dioxide is "fixed" from the atmosphere into the tissues of a leaf of lettuce through photosynthesis, the process through which green plants create simple sugars from the inputs of sunlight, carbon dioxide, and water. The leaf of lettuce finds its way into a salad, is eaten, and is then digested in the human body.

In the form of a simple carbohydrate, the carbon is then combined with oxygen (or "burned") at the cellular level to carry out a cellular function, converted into a molecule of carbon dioxide, and exhaled from the body.

Once returned to the atmosphere, the molecule of carbon dioxide is likely to reside there for many years. On average, more than half of the carbon dioxide released each year from all sources remains in the atmosphere many decades later, with 20% to 35% remaining for at least 200 years [4]. For example, much of the massive amounts of carbon dioxide released during World War II still remains in the atmosphere today, where it continues to drive the greenhouse effect year after year.

Over long periods of time, most atmospheric carbon is dissolved into the oceans and is fixed through photosynthesis into aquatic plants such as algae. As these plants, along with the animal populations that consume them, die and decompose, large quantities of carbon are deposited in ocean sediments. Over millions of years, these sediments sink farther into the Earth's crust and, in response to compression and heating from the planet's core, are transformed into oil, gas, and, in regions once occupied by swamplands, coal deposits. Therefore, the release of carbon back into the atmosphere through the burning of fossil fuels can be understood to constitute an additional pathway in the global carbon cycle.

However, the rate at which human activities are returning fossilized carbon to the atmosphere is far greater than the rate at which global oceans can absorb this excess carbon. This point is central to understanding the role of human activities in creating an *enhanced* greenhouse effect. Through the burning of oil, gas, coal, and forests, we have created an imbalance in the global carbon cycle, through which carbon is accumulating more rapidly in the atmosphere than it is being removed through absorption into the oceans. It is precisely this phenomenon that is measured by Keeling's curve, which has recorded since the late 1950s a steady rise in the atmospheric concentration of CO_2 from 315 parts per million (ppm) to today's 392 ppm. With each passing year, through the burning of fossil fuels, we return to the atmosphere a quantity of ancient carbon that requires hundreds of years to be released through natural processes. In so doing, we overwhelm the capacity of global oceans to regulate the level of carbon dioxide in the atmosphere, giving rise to an excess stock of atmospheric carbon and an enhanced greenhouse effect.

Reconstructions of climatic conditions over the past million years yield important insights into imbalances in the global carbon cycle. Through the analysis of ancient ice extracted from ice sheets miles thick atop Greenland and Antarctica, scientists are able to extract two key pieces of information. First, as snow accumulates in these regions, tiny air bubbles are trapped in the annual layers of ice. Entombed in these air bubbles are direct physical samples of ancient atmospheres, stretching back hundreds of thousands of years and recording changing levels of carbon dioxide.

A second useful piece of information is recorded in the chemical composition of the ice itself. Due to the presence of isotopes – for example, atoms of hydrogen that are characterized by the same number of protons as all hydrogen atoms but by a different number of neutrons – not all water molecules are the same. "Heavy" water results from the combination of two atoms of deuterium, a hydrogen isotope, with one atom of oxygen and exhibits a different pattern of ice formation in response to temperature changes than the non-isotopic form of water. As a result of this physical distinction, analysis of the isotopic properties of ice yields information on local temperatures at the time of ice formation.

Through the drilling of ice cores – narrow cylinders of ice extracted vertically from long frozen ice sheets – scientists can reconstruct a record of carbon dioxide and temperature trends reaching back to ancient millennia. As described by Kolbert (2006),

> [T]he Greenland ice sheet is made up entirely of accumulated snow. The most recent layers are thick and airy, while the older layers are thin and dense, which means that to drill through the ice is to descend backward in time, at first gradually, and then much more rapidly. A hundred and thirty-eight feet down there is snow that fell during the time of the American Civil War; 2,500 feet down, snow from the time of the Peloponnesian Wars, and, 5,350 feet down, snow from the days when cave painters of Lascaux were slaughtering bison. At the very bottom, 10,000 feet down, there is snow that fell on central Greenland before the start of the last ice age, more than a hundred thousand years ago. [5]

Such reconstructions have yielded further confirmation of the close association between atmospheric levels of CO_2 and temperature. As illustrated in Figure 1.3, CO_2 and temperature have exhibited a strong correlation over time, with periods of relative warmth accompanied by high levels of CO_2 and periods of relative cold accompanied by low levels of CO_2. The close correspondence of these trends – reconstructed from separate indicators in the climate record and confirmed by the analysis of thousands of ice cores in different regions of the planet – is remarkable. And it is also worrisome. At present, atmospheric CO_2, greatly augmented

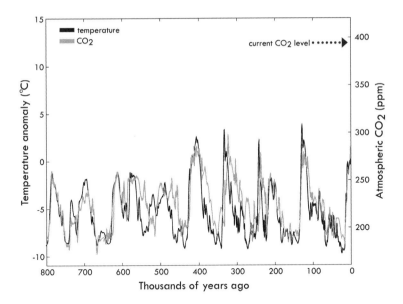

Figure 1.3   Historical temperature and CO_2 trends. *Source:* Adapted from
L. McInnes, Temperature and CO2 records. For information on source
data, see http://en.wikipedia.org/wiki/File:Co2-temperature-plot.svg.

by the transfer of billions of tons of carbon to the atmosphere through
the burning of forests and fossil fuels, has climbed to levels about 30%
greater than the highest concentrations revealed in 800,000 years of ice
core data [6]. Scientists cannot assert with 100% certainty that temperatures will continue to increase with rising levels of CO_2, but a divergence from this trend would mark a departure from a measured pattern almost 1 million years in the making. The more likely outcome, it
would seem, and the outcome strongly supported by a wealth of temperature measurements over the past few decades, is that the pattern is
holding.

An admittedly dire model for what has been characterized as a "runaway"
greenhouse effect is the planet Venus. Long known to exhibit surface
temperatures in excess of 800°F, Venus was once believed to be encircled
by a dense atmosphere of heat-trapping dust or particles. An unmanned
NASA mission to Venus in the late 1970s would reveal that about 97%
of the planet's atmosphere is actually composed of carbon dioxide. As a
result, virtually all the heat emitted from the planet's surface is trapped
by the atmosphere, creating a supercharged greenhouse effect. Although
there is little chance that the Earth's atmosphere could reach comparable

concentrations of CO_2, because even burning the entire global stock of fossil energy would not create such an imbalance, the study of Venus's atmosphere has yielded important insights for our understanding of an enhanced greenhouse effect on Earth.

Dr. James Hansen, a scientist with NASA and perhaps the world's most prominent expert on climate change today, would apply modeling tools developed through his doctoral work on Venus in the 1960s to the study of the greenhouse effect on Earth. At that time, the natural mechanism of the greenhouse effect was well established in the realm of climate science. Since the publication of Arrhenius's work on CO_2 and climate, little more than the name used to describe the phenomenon – with a widely read climate text of the 1930s proposing the phrase "greenhouse effect" as a more modern version of Arrhenius's "hothouse" metaphor – had changed.

Yet, although Keeling's annual CO_2 observations continued their inexorable march upward in the 1960s, there was little evidence from the global temperature record to suggest a warming trend. To the contrary, the decade of the 1960s was a period of relative cooling, now believed to be a product of heavy, sunlight-reflecting aerosol pollution in the upper atmosphere from the uncontrolled emissions of fossil-fuel use. However, Hansen and others recognized the potential for rising levels of CO_2 to eventually create a warming influence more powerful than countervailing climate drivers, and the rapidly advancing capacity of computers presented the best means of simulating such trends.

Working for NASA's Goddard Institute for Space Studies (GISS) in the early 1970s, Hansen was part of a team of scientists attempting to use newly available weather satellite data to assess the implications of rising carbon dioxide concentrations for long-term climate change on Earth. The group sought to develop a computer model capable of accounting for the wide array of environmental influences on climate over time, such as changes in the receipt of solar radiation, the changes in cloud formation, the distribution of heat energy through ocean and atmospheric circulation, and, of course, the influence of greenhouse gas emissions on temperature, among a very large set of additional climate drivers. To be useful in projecting future climate in response to changes in these drivers, such a global climate model would need to be effective in reproducing climate conditions of the past, for which direct observations are available to validate the model. It is fair to say that the earliest models had only limited success in reproducing past climate episodes.

However, model accuracy would improve over time. Most important, two-dimensional models of the 1960s would be replaced by three-dimensional models in the 1970s, enabling researchers to capture the transfer of energy and moisture between different levels of the atmosphere. Soon, research centers such as NASA's GISS, the Geophysical Fluid Dynamics Laboratory in New Jersey, and the National Center for Atmospheric Research in Colorado would employ teams of scientists tasked with developing detailed algorithms for specific components of the global climate system to feed into these general circulation models (GCMs). By the 1980s, these models were demonstrating increasing proficiency in reproducing past climate episodes and were providing a reasonable window into future climates characterized by rising levels of greenhouse gases in the atmosphere. A decade later, reproductions of historical climate trends generated by the most widely used GCMs were shown to have a greater than 0.9 correlation with actual surface temperature observations – reproducing these trends with a surprisingly high degree of accuracy [7].

Buoyed by the increasing precision of his climate models, in 1988 Hansen would assert in testimony before the U.S. Congress with "99 percent" confidence that human activities were altering the global climate, stating that the "greenhouse effect has been detected, and it is changing our climate now" [8]. He went on to suggest without equivocation that human actions had reached the point of "loading the climate dice," and he urged Congress to take action to curb the rate of greenhouse gas emissions.

The events of that year had produced an unusually receptive audience for Hansen's analysis. On only the second day of summer, temperatures outside the committee room in which Hansen was testifying had soared to 101°F, one of 21 consecutive days in Washington, DC, with temperatures in excess of 90°F. That summer, more than 2,000 daily high-temperature records would be broken across the United States, while a prolonged drought plagued the nation's interior. As described by one Montana rancher who had lived through the Dust Bowl years of the 1930s, "I've never seen it like this, not even in '34. My God, look at it. It's as bare as a dance floor" [2]. By its end, 1988 would prove to be the hottest year ever recorded.

Hansen's predictions would be afforded greater credibility over the decade following his congressional testimony. Convinced of the increasing influence of rapidly climbing greenhouse-gas concentrations on climate, Hansen offered a simple wager to any of his colleagues interested in challenging the mounting evidence of a warming planet. To any willing taker, Hansen would pay $100 if one of the first three years of the 1990s did not set a new record for the highest average global temperature ever measured. Despite portrayals in the media, then and now, of a heated debate on the issue within the scientific community, only a single colleague accepted Hansen's challenge. The suspense lasted for only six months – 1990, the first year eligible to break the all-time temperature record, was found to be the hottest in 110 years of record keeping.

In the years since 1990, no scientist has stepped forward to double-down on Hansen's bet – and with good reason. For, although computer models provide a critical tool to understand how climate may change many years into the future, one needs only a backyard thermometer to discern a sharp warming signal in the present period. The year that Hansen testified to the U.S. Congress on the certainty of human-induced climate change – 1988 – was found not only to be the hottest year ever recorded but it also surpassed the previous highest temperature anomaly – set in 1987 – by a healthy margin. When measured as a deviation from a long-term trend, the 1988 temperature anomaly was about 15% greater than the preceding year and more than 60% greater than the record of any prior decade. In short, 1988 not only was hot: it was off the charts in terms of historical extremes, including the Dust Bowl years of the 1930s. Yet, the 1990s would render the '88 record almost trivial.

If Hansen had found takers for his wager after 1990, he would have won the bet another five times. Six of ten years in the decade of the 1990s surpassed the 1988 record: 1990, 1991, 1995, 1997, 1998, and 1999 were all hotter. The 1990 anomaly was another 15% greater than the 1988 record; the 1998 anomaly, the hottest year of the 1990s, was an astounding 80% higher than 1988. Evidence of the extremity of global warming already underway abounded in the 1990s.

More revealing of an underlying trend than the temperature extremes in the 1990s was the pace at which records were falling. In the period between 1900 and 1988, new temperature records were established on average every 12.6 years. Between 1988 and 1998, that interval had decreased to 2.5 years. This fact alone is worthy of restatement: new global temperature extremes, requiring in recent history an average of 13 years between occurrences, were in the 1990s being experienced every 30 months – recurring with greater frequency than the Olympic Games. Yet, despite this feat, the first decade of the 21st century would surpass the extremity of the 1990s. Of the ten hottest years ever recorded by human instruments, *nine* occurred between 2001 and 2010. Whereas not a single year had approached the extreme benchmark set in 1988 in the preceding 108 years of record keeping, every single year between 2001 and 2010 would exceed this outlier by a wide margin. The greatest temperature anomaly ever recorded (at the time of this writing), in 2010, was *more than double* that of the 1988 extreme and 10% greater than the previous decade's record.

The statistical probability that such a string of excessively hot years – occurring not only with increasingly frequency but also with accelerating intensity – could result absent an underlying climatic shift is effectively zero. The implications of these trends should be apparent to any sentient person alive today: the Earth's climate is changing, and basic physics tells us that the long-term stability of an earlier climate will not return during the lifetime of anyone reading these pages. As Hansen elegantly summarized in his testimony to Congress, now almost a quarter-century ago, the injection of massive amounts of carbon dioxide into the atmosphere has loaded the climate dice. Due to the unalterable physical principles linking greenhouse gases to the trapping of heat, and the long residence time of CO_2 in the atmosphere, we are already committed to many decades of continued warming – to a world in which every coming year has better than even odds of being hotter than the last. To wager on continued warming in such a world is to wager with the mathematical advantages of the house, the eventual outcome never in doubt.

In the scientific community today, the work of a climate skeptic is an increasingly lonely enterprise. Although limitations in computer models and shortcomings in available measurements gave rise to a few competing theories to explain observed warming trends in the last decades of the 20th century, none of these theories, individually or in concert, has proven sufficient to account for the pace and intensity of rising temperatures. Absent the influence of human activities, the dramatic changes in climate experienced to date simply cannot be explained. To arrive at this conclusion is not to discount the importance of natural processes to changes in global climate but rather to acknowledge what a massive compilation of global measurements has demonstrated: human activities are now the dominant driver of long-term climate trends.

The global scientific community has asserted a human role in climate change as a broad consensus, first in 1995, based on an exhaustive analysis of natural and human drivers of climate change, known technically as climate "forcings." Throughout the decade of the 1980s and into the 1990s, a few important inconsistencies in climate datasets raised reasonable questions about the extent to which greenhouse gas emissions were driving observed warming trends. Because natural climate forcings, such as variation in the intensity of the Sun's output or cyclical patterns in the warming of the ocean's surface, are also known to influence global temperatures, direct measurements of these forcings have now been made and compared with the influence of greenhouse gas emissions. The results of these analyses leave little doubt as to the principal driver of the rapid warming.

To remain skeptical of the consensus view of the global scientific community today is to subscribe to one of three general viewpoints: (1) the Earth is not warming; (2) the Earth is warming primarily through natural forcings; or (3) the Earth is warming primarily through human-induced forcings, but natural feedback mechanisms will curtail this warming. It is worthwhile to examine each of these views in turn.

The Earth is not warming. Of all the competing theories to a human-enhanced greenhouse effect, the idea that the Earth is not warming can be dismissed most easily. No credible climate scientist is on record today as doubting the existence of a clear warming trend over the course of the last three decades. The measurement of such a trend by thousands of land- and sea-based meteorological stations across the planet confirms a strong and consistent warming pattern since the 1980s. However, before the late 1990s, temperature measurements derived from satellites showed a pattern of moderate cooling, providing a basis to question the true direction of global climate trends. Yet, it was discovered more than a decade ago that this inconsistency resulted from a failure to account for a gradual decay in the orbit of weather satellites over time. As these satellites gradually lose altitude, global temperatures recorded by onboard instrumentation appear to decrease. Measurement of the decay in satellite orbits has enabled this temperature record to be mathematically corrected, resulting in the confirmation of a warming trend by both land-based and satellite temperature records [9].

Observed warming trends are primarily attributable to natural forcings. Periodic shifts in global climate revealed in the paleo-climatic record attest to the existence of natural mechanisms that can lead to substantial global warming or cooling. For example, in addition to the role played by atmospheric composition in the cycles of glacial and interglacial periods, as proposed by Arrhenius, changes in the Earth's orbit around the Sun are also known to influence patterns of glaciations. Known as the Milankovitch cycles, changes in the shape of the Earth's orbital path and the position of the Earth's axis can increase or lessen the intensity of energy received from the Sun over different time scales. Also important is the 11-year solar cycle, through which changing sunspot activity increases the receipt of solar energy over an established cycle. In light of the measured and reconstructed influence of these cycles on climate over time, many scientists have questioned whether recent warming trends are more directly attributable to changes in the receipt of solar energy than to increasing greenhouse gases.

Measurements of the receipt of solar energy over the last several decades, a period in which mean global temperatures have increased substantially, indicate that the intensity of this radiation has remained largely constant, outside of fluctuations in the 11-year solar cycle. Yet even these fluctuations have not been found to exert a significant influence on global temperatures. During the years of 2005–8, for example, the intensity of solar output reached its lowest point since satellite measurements first became available in the 1970s. Yet, these years constitute three of the hottest years ever measured. In addition, recent warming trends have shown winters to be warming more rapidly than summers, and nighttime temperatures to be rising more rapidly than daytime temperatures, exactly the opposite of what one would expect to find were increasing solar activity responsible for global warming trends [10].

Global climate models offer an important tool for understanding the influence of solar cycles and other natural drivers on observed warming trends. Scientists make use of computer models for two basic purposes. The first and most widely understood purpose is to estimate how the Earth's climate will change in future years. If global climate models can be shown to reproduce past, observed climates with sufficient accuracy, these models provide a reasonably reliable basis to estimate how future climates will respond to changing environmental conditions. A second and equally important use of climate models is to disentangle the relative contribution of the many different drivers of global climate in present and past periods. Because global climate patterns are sensitive to a large number of natural and human-driven forcings, ranging from greenhouse gas emissions to changing solar patterns to changing patterns of glaciations, only the processing power offered by large clusters of computers provides a means of understanding how a change in any one input to the system influences the overall system.

The precision with which computer models can reproduce historical climate depends directly on the precision with which inputs to these models have been measured over time, and great advances have been made in those measurements. Over the period of at least three decades, aided in large part by the availability of satellite data, scientists have been able to accurately measure important inputs to climate models, such

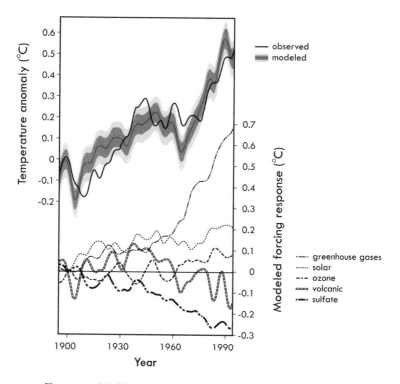

Figure 1.4   Modeled drivers of global temperature trends.
Source: Adapted from R. Rohde, Global Warming Art. For more
information on source data, see http://www.globalwarmingart.com/
wiki/File:Climate_Change_Attribution_png.

as the intensity of radiation received from the Sun, the concentration
of greenhouse gases in the atmosphere, and the percentage of polar
regions covered by ice. Accurate measurements of these conditions, combined with an increasingly comprehensive understanding of the physical
relationships between these variables, have improved the effectiveness
of global climate models in reproducing past climates and in measuring
the sensitivity of global climate to specific inputs. It is with the aid of
these models that scientists can test, with a high degree of confidence,
competing theories proposed to explain recent warming trends.

The competency with which the leading global climate models now
reproduce historical climate is clearly illustrated in Figure 1.4. The figure
presents the results from a global climate modeling study through which
modeled global temperature changes are compared to observed global
temperature changes. The model also assesses the influence of observed
changes in five natural and human drivers of climate: greenhouse gas

emissions, the intensity of solar radiation, levels of ozone in two layers of the Earth's atmosphere (the troposphere and stratosphere), sunlight-reflecting pollutants emitted through volcanic eruptions, and sunlight-reflecting pollutants emitted through fossil-fuel combustion.

As illustrated in the top half of the figure, global climate models are able to reproduce historical temperature observations with impressive accuracy over periods of a decade or longer. Although the modeled temperature may differ significantly from temperature observations in any single year, these models capture decadal temperature trends quite well. Overall trends include an extended period of warming from the early to mid-20th century, followed by a period of stable or declining temperatures during the 1950s and 1960s, and then a sharp period of warming through the 1980s and into the 1990s. In all periods of a decade or longer, the models represent accurately the direction of climate change, one of predominant warming or cooling, and the general rate of change, whether rapid or moderate. To any objective observer, these results attest to the success with which climate scientists have been able to reproduce extraordinarily complex climatic phenomena with today's global climate models.

The five fluctuating lines presented in the bottom half of the figure quantify the influence of distinct climate-forcing agents over the same period of time. Equipped with accurate measurements or reconstructions of these forcing agents and knowledge of their role in driving temperature change, scientists can use computer models to assess the relative influence of each forcing agent on climate. For example, given that the intensity of solar output received by the Earth has remained largely flat since the 1950s, there is little scientific basis to support the theory raised by some skeptics that changes in solar intensity account for most of the recent warming trend. Rather, marginal increases in solar output since the 1980s are found to account for less than a quarter of the increase in global temperatures during that time.

What is clear from a wide set of global climate model simulations is that no natural climate-forcing agent, alone or in concert with other such agents, is sufficient to account for the dramatic rise in temperatures presently underway. This increase in temperature is matched only by the unrelenting increase – year after year and over the period of decades – in atmospheric concentrations of greenhouse gases. It is for this reason that the global climate science community has reached the broad consensus that human activities are the dominant driver of ongoing changes in climate. And it is for this reason that the number of formally trained scientists promoting theories of natural change has dropped to little more than a handful of holdouts. The evidence in favor of human-induced warming is simply too great.

The Earth is warming through human-induced forcings, but natural feedback mechanisms will curtail this warming. Of all the challenges to the theory of a human-enhanced greenhouse effect, the potential for a change in the global ecosystem to occur in response to rising temperatures, counteracting a warming trend, is perhaps the most plausible. Indeed, the capacity for Earth to regulate its own climate in response to changing conditions is the very basis of the Gaia theory first proposed by British scientist James Lovelock in the 1960s. As understood by this theory, the Earth's plant and animal life – referred to collectively as the *biosphere* – has played an active role in maintaining environmental conditions favorable to its survival over millions of years. As global temperatures rise to dangerous levels, for example, the growth of plant life accelerates and, in so doing, draws down atmospheric levels of CO_2 through photosynthesis, limiting rising temperatures to a tolerable range. Likewise, excessive global cooling is believed to be offset through a reduction in photosynthetic activity, which raises atmospheric CO_2 and enhances the greenhouse effect.

In a similar vein to the Gaia theory, some scientists have posited that a prolonged warming trend brought about through greenhouse gas emissions could be offset through a natural cooling mechanism, referred to technically as a "negative feedback." One such potential negative feedback is cloud formation. For example, Dr. Richard Lindzen, a climate scientist at the Massachusetts Institute of Technology and a prominent critic of climate-change projections produced by the Intergovernmental Panel on Climate Change (IPCC) – the international scientific body convened by the United Nations to develop consensus science on climate change – has hypothesized that rising sea-surface temperatures in the tropics will lead to a reduction in high-altitude cirrus clouds. Such a reduction, he predicts, will permit a greater proportion of terrestrial radiation to pass out of the Earth's atmosphere, offsetting enhanced warming brought about by rising greenhouse-gas concentrations.

Since the publication of Lindzen's hypothesis in 2001, few studies have found evidence of either a reduction in cirrus-cloud formation or of a cooling effect attributable to such a change. Most studies have found clouds to play a dual role in climate change, in that the suspended water droplets of clouds are capable of both absorbing outgoing terrestrial radiation and reflecting away incoming solar radiation. In fact, it is estimated that without the natural envelope of clouds, covering more than half the planet at any particular time, the Earth would be more than 20 degrees hotter. A specific test of the Lindzen hypothesis with the aid of satellite measurements of cloud formation and temperature found that a reduction in these clouds is associated more strongly with a warming effect than a cooling effect [11].

Lindzen's hypothesis highlights an increasingly problematic fact for climate change skeptics. After more than 30 years of a sustained and rapid warming trend, there is very little evidence that any known or theorized feedback mechanisms are developing to offset an enhanced global greenhouse effect. Whether it be in the form of increased absorption of CO_2 by the world's oceans or forests, changes in cloud formation, or increased reflectivity from growing aerosol pollution, none of these or other theorized effects has been demonstrated to be significantly counteracting warming brought about through greenhouse gases. To the contrary, positive feedbacks in the form of rising levels of water vapor in the atmosphere and a loss of sunlight-reflecting ice at the poles have been clearly demonstrated to be amplifying the greenhouse effect, with the net effect evidenced by rapidly increasing global temperatures.

Despite the overwhelming evidence in support of a warming planet, confirmable by anyone with access to a thermometer in virtually any reach of the planet, new life was breathed into all brands of climate skepticism in 2009 through what would become ingloriously labeled as "climategate." In the days preceding an international meeting on climate change in Copenhagen, a set of email files from a single research center in the United Kingdom were hacked into and distributed to the media. Included in these materials were thousands of emails exchanged between researchers at the Climate Research Unit of the University of East Anglia and other climate scientists over a period of 13 years; a handful of these emails were seized upon by skeptics as revealing attempts to manipulate historical climate data.

The most damning emails, as portrayed in the media, concerned a long-standing inconsistency in historical tree-ring data, which have been used as a proxy for estimating climate trends predating the instrument record. Although the tree-ring temperature estimates approximate those of ice core data and other proxies over several hundred years, by the middle of the 20th century, these estimates suggested a leveling off of temperature trends, despite the direct evidence of warming in recent decades provided by thermometers. No consensus explanation for this recent divergence yet exists, but one plausible reason is that the growth patterns of trees were themselves affected by climate change during the 20th century. Climate scientists have debated the best means of presenting these data so as not to suggest a recent cooling trend at odds with the observed temperature record. As noted in rather inartful shorthand by one of the scientists in an email exchange, there was a

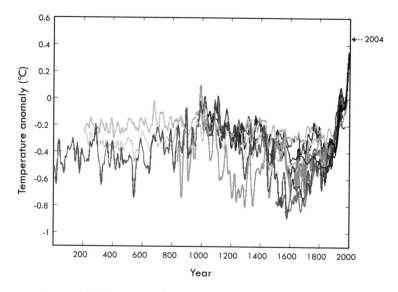

Figure 1.5   Reconstructed global temperature trends. Each curve represents a different temperature proxy; black line represents temperature observations. *Source:* Adapted from R. Rohde, Global Warming Art. For more information on source data, see http://www.globalwarmingart.com/wiki/File:2000_Year_Temperature_Comparison_png. (*A color version of this figure can be found in the color plate section after page 86.*)

perceived need to "hide the decline" in the tree-ring temperature record so as not to create confusion.

Some viewed this email exchange as clear evidence of an attempt to conceal compelling evidence at odds with the theory of an enhanced greenhouse effect. However, an important fact not referenced in most media accounts of the emails suggests otherwise: the inconsistency in tree-ring data was a widely recognized issue in the science community, among proponents and skeptics alike, and had been the focus of prominent peer-reviewed articles in the literature (e.g., [12, 13]). It is precisely because of such imperfections in proxy temperature measurements that multiple proxies are used to reconstruct historical climate trends, ranging from tree rings to ice-core data to sediment-core data to measurements of coral growth. As illustrated in Figure 1.5, these many proxies, when viewed in concert, suggest a general pattern of climate change over time and are clearly indicative of warming in the 20th century.

What the climategate episode most clearly attests to, if anything of use, is the deeply flawed reasoning on which many climate skeptic arguments are grounded. To believe that the climategate emails uncover a widespread attempt by prominent climate scientists to manipulate their data is to believe that the global science community is complicit in a collective effort to promote a uniform theory of human-driven climate change. It is to assume that the surest path to advancement in scientific careers is to parrot the arguments and findings of one's peers. Yet, in what professional career path is conformity prerequisite to the attainment of national and international recognition? Nobel laureates become so through demonstrating what no one else has demonstrated; Galileo, Newton, and Einstein are not universally recognized names because of their confirmation of other's findings. There is likely no surer path to recognition in the realm of climate research today than to present compelling data that the Earth is, in fact, not warming due to human actions. That the community of climate skeptics remains remarkably small in number today and tellingly absent from the peer-reviewed literature is a testament not to a conspiracy of thought but rather to an absence of evidence.

If any data were inappropriately modified by the handful of scientists involved in this episode, a conclusion unsupported by the institutional inquiries following the event, it is far more likely that such actions were born of fear than ambition. As anyone genuinely steeped in climate research today can attest, the overall trends are unambiguous and they are troubling. Scientists privy to these data are justified in wondering if the muted language required by the medium of scientific publication is proving sufficient to avert a profound loss of human life, even in the span of their own lifetime. And so an inclination to elevate one's volume in response to an ever-increasing threat, in the face of an increasingly unhearing public, is not surprising. Yet, the content of the message remains unchanged and inalterable: the evidence in support of human-induced climate change is overwhelming.

That a small and largely uncredentialed community of climate skeptics has succeeded in raising doubts among the general public regarding the veracity of a human-enhanced greenhouse effect, and has done so in the face of the most extreme warming trend experienced in human history, is a truly remarkable feat. When 9 of the last 10 years (and 17 of the last 20) have been the hottest ever measured, how is it possible to convince anyone living through this period that there is no apparent warming trend – that the likelihood of a temperature decrease in the next year is equal to the likelihood of a temperature increase?

One may counter that most skeptics do not deny a warming trend, only its provenance in human activities, and this may be a fair argument, albeit one unsupported by scientific observation. Yet, in the face of such an extreme shift in climate, rendered plainly manifest in the instrument record, would it not be advisable to prepare for additional warming? For, whether the ocean is rising due to acts of humans or acts of God, the ocean is indeed rising, glaciers are melting, droughts are spreading, and storms are growing more intense. Yet, at the start of the second decade of the 21st century, more than 114 years since the potential for humans to alter the Earth's climate was mathematically demonstrated, 52 years since the first measurements of rising CO_2 levels were recorded, and almost a quarter-century since the nation's top climate scientist testified as to the clear signal of human-driven warming to the U.S. Congress, the United States has not passed a single bill or allocated a single dollar to prepare for the most profoundly threatening environmental conditions ever anticipated by science.

The moment we now occupy is, it would seem, unprecedented.

For the most vulnerable areas of the planet, the century's massive climate migration has already begun. To see this process in real time, one need only search an Internet map server for the Island of Lohachara, in India's Bay of Bengal, and toggle between a map of the island and a recent satellite image. Depicted on the map will be an island of about three miles square, once inhabited by 10,000 residents, who for generations farmed the small island and lived off the sea. Found at the same location on the satellite image will be a barely visible land mass, just beneath the surface of the water. Clearly discernible through the water is the outline of a heavily eroded roadway, the last remaining evidence of long term human settlement. On the map, the island's boundary remains, but the name has been removed, its geographic significance as a land feature now gone.

The rising sea levels that have already swallowed Lohachara and numerous other low-lying islands are not a theory but a manifestation of the global shift in climate that is currently underway. To understand how this shift will affect the planet over the coming decades, scientists must rely on global climate models to project the implications of rising global temperatures for a range of interdependent environmental systems. Although such models have been demonstrated to reproduce historical climates with an impressive degree of accuracy, they can only project future conditions within a range of certainty. Yet, at both the high and low ends of this range, the direction of projected impacts is consistently unfavorable to human and environmental well-being.

Figure 1.6 presents projections of future temperature change developed for the period between the present day and 2100 by the IPCC. The colored trend lines depict the expected ranges of temperature change

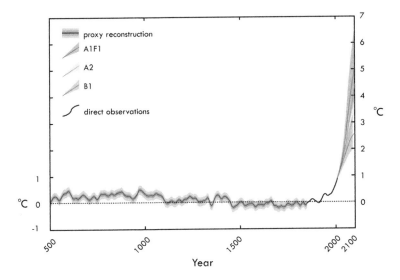

Figure 1.6   IPCC global temperature projections. *Source:* Adapted from I. Allison, N. Bindoff, R. Bindschadler, et al., *Copenhagen diagnosis: Updating the world on the latest climate science*, University of New South Wales Climate Change Center, 2009. (*A color version of this figure can be found in the color plate section after page 86.*)

in response to specific modeling scenarios, each with differing assumptions pertaining to the level of greenhouse gas emissions in future years. The "B1" growth scenario makes the most optimistic assumptions about emissions reductions, whereas the "A1F1" growth scenario assumes the continuation of historical growth rates in emissions (i.e., "business as usual"). The "A2" scenario falls between these two extremes. The black line shows trends during the period for which we have direct measurements of global temperatures from thermometers and satellites, and the purple line represents the reconstruction of trends that predate the instrument record through the use of ice-core data, tree-ring data, and other proxies. Based on these model simulations, the IPCC projects in its most recent assessment report (2007) that by the year 2100, mean global temperatures will increase between about 2 and 12°F [14].

These results show that even at the lowest end of the modeled range, the projected increase in temperatures is greater than that revealed in 1,500 years of reconstructed temperatures and likely not experienced for hundreds of thousands of years. As estimated by Arrhenius more than a century ago and confirmed by recent estimates, an increase in average global temperatures of 7°F, about the midpoint of these projections, is equivalent to the magnitude of warming between the last ice age and today, representing massive changes in climate. In addition to rising temperatures, clear evidence of a rapidly changing climate can be found in any global environmental system, ranging from the oceans to the land surface to the atmosphere.

The loss to rising seas of Lohachara and other small islands marks only the beginning stages of a human drama that will play out over the present century and beyond. Increasing temperatures influence global sea levels through two mechanisms. First, the melting of ice sheets atop Antarctica and Greenland, combined with the shrinking of glaciers around the planet, is discharging massive amounts of water into the world's oceans, contributing to a gradual rise in sea level. Second, the volume of water expands with the addition of heat, contributing to a rise in sea levels independent of the melting of ice sheets. Due to the extensive volume of the global oceans, many decades will be required for the heat accumulating in the atmosphere to be fully distributed through the depths of the oceans. As a result, the oceans are on course to continue rising for many decades independent of future emissions of greenhouse gases.

Figure 1.7 presents sea-level change relative to 1990 as measured by tide gauges and satellite observations. As illustrated, sea levels have been rising sharply since the mid-1980s, matching closely the worst-case scenario as modeled by projections from the global scientific community. Since 1993, the first year in which reliable satellite measurements of sea level became available, the oceans have been rising at a rate of 3.4 mm per year – about 80% higher than the IPCC's best estimate for this period [10]. This rapid rate of increase suggests that the IPCC's current projections for sea-level rise throughout the present century, between about 6 inches and 2 feet, are also too low. The most recent projections suggest rates of sea-level change much greater than these estimates, enough to displace more than 160 million people living in low-lying coastal areas. With a possible rise of more than 6 feet this century [10], low-lying coastal areas, including large cities such as New Orleans, are in danger of being lost before the century's end [15].

A tendency for recent observations of environmental change to exceed previously modeled projections has become a common theme in the climate-change literature. Because scientists are conservative by training, model simulations are based on only the most strongly supported assumptions, which may tend to underemphasize important agents of climate forcing. In addition, scientists average the results of numerous climate model simulations to minimize statistical error, a

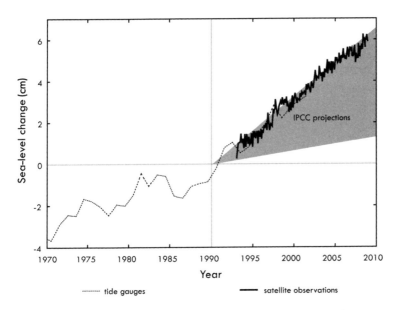

Figure 1.7   Observed sea-level change. *Source:* Adapted from I. Allison, N. Bindoff, R. Bindschadler, et al., *Copenhagen diagnosis: Updating the world on the latest climate science*, University of New South Wales Climate Change Center, 2009.

practice that has the desired effect of limiting unusually high or low projections. Although often portrayed by skeptics as representing extremist views, mainstream climate scientists are compiling a consistent track record of significantly underestimating the true pace and extent of climate change.

A key reason why sea-level rise has been underestimated pertains to the rate at which polar regions are melting. Figure 1.8 illustrates the rate at which sea ice is disappearing from the Arctic in late summer. The modeled extent of sea ice across the Arctic over time is measured by the black line, with the dashed lines showing the range of modeled outcomes. The gray line illustrates the extent of sea ice in late summer as measured by satellites and reveals how much faster sea ice is melting annually than projected by model simulations. Found in 2007 to be about 40% lower than average, the extent of Arctic sea ice is now retreating at a rate of about 11% per decade and may seasonally disappear altogether in another few decades [10]. Such an outcome is problematic not only for rising sea levels but also for the pace of global warming. This is because the highly reflective nature of ice and snow – a property known as "albedo" – helps regulate global temperatures by reflecting away incoming solar

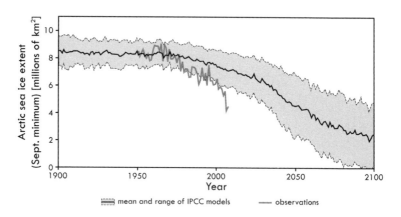

Figure 1.8   Observed and projected extent of Arctic sea ice. *Source:* Adapted from I. Allison, N. Bindoff, R. Bindschadler, et al., *Copenhagen diagnosis: Updating the world on the latest climate science*, University of New South Wales Climate Change Center, 2009.

radiation received in the polar regions of the planet. The melting of ice sheets therefore not only raises sea levels but also changes the color and reflectivity of expansive polar zones. As explained by one climate scientist,

> Not only is the albedo of the snow-covered ice high; it's the highest of anything we find on Earth. And not only is the albedo of water low; it's pretty much as low as anything you can find on Earth. So what you're doing is you're replacing the best reflector with the worst reflector [5].

The end result of this process is an amplification of the rate of warming brought about through a response to initial warming, or a positive feedback. And this is precisely why early action is required to slow the pace of climate change: once we start to see the obvious effects of an enhanced greenhouse effect, greenhouse gases are no longer the only driver of continued warming. Mitigation at this point requires aggressive action and on more than a single front.

Rising temperatures in the world's oceans also hold many implications for the occurrence of extreme weather events. The most direct of these effects can be seen in the growing intensity of hurricane activity. Hurricanes are responsive to the temperature of the sea surface, which is why late summer, when these temperatures are highest, is the period of greatest hurricane activity. The latest research finds that the frequency of the most destructive hurricanes has increased in recent decades [16, 17].

Based on an analysis of satellite data collected since 1980, an increase in mean global temperatures of 1.8°F has been found to correspond to a 30% increase in the number of Category 4 and 5 hurricanes [17]. Thus, a 5- to 6-degree rise in global temperatures, well within the range projected for this century, may be expected to produce an approximate doubling in the number of the most destructive hurricanes per year.

Rising sea-surface temperatures also influence the number and intensity of storm events over land. As demonstrated by the increase in storm events associated with the El Niño Southern Oscillation – a cyclical warming of sea-surface temperatures in the Pacific Ocean that occurs every two to seven years – the temperature of the oceans holds direct implications for the extent and intensity of precipitation over land. As temperatures climb over water and land as a product of the global greenhouse effect, more moisture is retained by the atmosphere, giving rise to more intense storm events. A clear trend in the number of heavy rain and snowfall events is illustrated in Figure 1.9, where the bars show the percentage of the United States subject to a higher than average number of days per year with extreme precipitation events (the trend line represents the five-year moving average of these events).

Of course, the most threatening aspects of a warming planet, as made clear by the European heat wave of 2003, are likely to result from the intensifying heat itself. At the low end of the IPCC temperature projections, a steadily rising rate of human mortality is expected in response to extensive heat waves, droughts, intensifying air pollution, and more destructive storm events. In addition, it is estimated that 30% of the planet's species may be driven to extinction by a 2 to 4°F rise in global temperatures. In other words, under the most optimistic scenario envisioned by the global scientific community – that associated with aggressive global mitigation efforts, the early stages of which are today nowhere in evidence – roughly one-third of all animal and plant species on Earth could be lost over the course of this century alone [18].

At the high end of this range – an increase in mean global temperatures of about 12°F – the impacts on human and ecological systems approach the unimaginable, with a sharp reduction in the global food supply a virtual certainty. Although few scientists have been willing to speculate in print on the impacts of the IPCC's worst-case scenario, there is evidence from the geological record to suggest very dire consequences indeed. The last time global temperatures climbed this high and so precipitously, about 250 million years ago, as much as 95% of the world's species were driven to extinction, an outcome associated with a runaway greenhouse effect [19, 20].

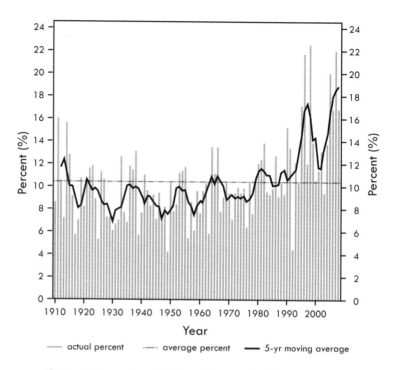

Figure 1.9   Percentage of United States experiencing extreme precipitation events. *Source:* Adapted from I. Allison, N. Bindoff, R. Bindschadler, et al., *Copenhagen diagnosis: Updating the world on the latest climate science*, University of New South Wales Climate Change Center, 2009 (data updated from K. Gleason, J. Lawrimore, D. Levinson, et al., A revised U.S. climate extremes index, *Journal of Climate*, 21, 2124–2137, 2008).

How likely is this worst-case warming scenario? Trends in global emissions of greenhouse gases over the last few decades provide the only direct basis for projecting future climate change. In the late 1980s, at the time of James Hansen's testimony to the U.S. Congress on the certainty of human-induced climate change, global emissions of CO_2 were increasing at a rate of about 1% a year. Over the course of the following two decades, a period in which evidence in support of a human-enhanced greenhouse effect grew increasingly more robust each year, and a period in which an international treaty to slow the growth in heat-trapping emissions was adopted, global emissions of carbon dioxide increased threefold, to more than 3% a year [21]. At this present rate of growth in emissions, the planet is on course to realize the IPCC's worst-case scenario for climate change within the lifetime of a child born today.

However dire projections of future warming appear to be, and how-ever compelling the underlying science, few of us have been moved to make meaningful changes to slow ongoing climate change. Although some of the responsibility for inaction can be laid at the feet of the climate skeptics and denialists, a more significant impediment to behav-ioral change is the wholesale decoupling of climate change from the geography and timescales of our own lives. For, however threatening a precipitous rise in mean global temperatures may be in comparison to historical trends, no one reading these pages lives at the global scale. And few of us anticipate living to see the year 2100, the year to which almost every major scientific projection directs our attention. Indeed, it is in the decoupling of climate change from meaningful space and time that the climate-change denialists find their most potent weapon, gleefully seizing on any aberration from a long-term warming trend in the places we live to prop up their otherwise unsupportable arguments.

That climate scientists have emphasized the global and the long term over the local and the immediate is not surprising given the scale of the physical mechanisms at work and the risk of establishing short-term markers that, if disproven, could undermine the credibility of their pro-jections. In this sense, an averaging of projected impacts over the planet as a whole and over the period of centuries permits climate scientists to emphasize general trends rather than specific occurrences that are much harder to predict with confidence. But the practical effect of this averaging is to divorce climate phenomena from the reality of people's lives and the urgency of their own circumstances.

As I show in the following chapters, an emphasis on the global and the long term has not only numbed the general public to the dangers of climate change but has also undermined the strength of the scientific case itself – a case that only grows more compelling when considered in context.

2
_____

The Climate Barrier

Six years before Europe's catastrophic heat wave, a hard freeze in southern Florida would reveal as much about the planet's changing climate as
the extremity of the 2003 disaster. Although no one is reported to have
perished in the winter storm, much of Florida's citrus crop was lost when
subfreezing conditions persisted for two days in January 1997. In total,
the crop losses – affecting not only oranges but also a wide range of winter vegetables – were estimated at more than $300 million, with prices
for some produce rising by more than 80% and about 100,000 seasonal
farm workers losing their winter employment [1, 2]. Long a hazard of the
orange-growing industry, winter cold snaps are to be expected from time
to time. But the extremity of this particular freeze was unlikely an act of
Mother Nature alone.

That so much of the state's orange crop was clustered in southern Florida at the time of the storm was itself a byproduct of occasional
freezes and crop losses in the preceding decades. In most respects, the
northern half of the state provides superior growing conditions for the
orange industry. The well-drained, sandy soils of northern Florida have
traditionally required lower land-preparation costs and produced superior crop yields than the water-saturated and more disease-prone lands of
southern Florida. As a result, virtually all orange growing by the mid-20th
century was situated in the north-central portion of the state, around
Orlando. Yet, the impacts of a few severe freezes in the early 1980s,
culminating in the estimated loss of one of every three orange trees,
would initiate a widespread migration to the southern reaches of the
state where the historical frequency of such freezes was far lower. Over
the decade of the 1980s, the five counties bordering Lake Okeechobee
in the southern half of the state experienced an almost 90% increase in
the number of orange trees, while a cluster of counties around Orlando
witnessed a comparable decline [3].

Relocating much of the orange industry to southern Florida was no small task. Before the 20th century, almost half of the land area of southern Florida was marsh and other wetlands unsuitable for crop production. To expand the acreage of land available for both agriculture and urban development, the U.S. Corps of Engineers undertook one of the most extensive water drainage and diversion projects in human history. Through the development of a system of levees, artificial channels, and pumping stations, 50% of the original area of the Everglades was drained to create new land for agriculture and urban growth. The federally sponsored project was a welcome boon to a citrus industry looking for warmer climes. However, the massively reengineered landscape would come at an unexpected cost: changes to the region's land surface would also bring about changes to its climate.

Evidence of the influence of the Earth's land surface on climate is easy to find in our everyday lives. For anyone who has walked barefoot through wet grass or stepped gingerly across sun-heated asphalt, the direct effects of land-surface properties on temperature are readily apparent. Although the influence on regional climates of a single lawn or parking lot is unquestionably small, as the spatial extent of land-surface changes grows larger, the magnitude of the climatic influence increases as well.

The profound influence of the Earth's surface on climate is most clearly manifested in regions with expansive areas of a single land-cover type. Desert regions, for example, experience relatively high temperatures during the daylight hours due to the absence of vegetative cover for shading and the low availability of soil moisture for evaporation, which cools the air. Endowed with vast quantities of water, tropical rainforests are far cooler than most desert environments during the day, despite their greater receipt of solar energy at their equatorial latitudes. In both extremes, the prevailing environmental conditions are not a product of regional differences in greenhouse gas concentrations but rather reflect the strong influence of land-surface properties on climate.

Long interested in the influence of landscape change on climate, a team of researchers at Colorado State University and NASA surmised that the extensive draining of wetlands in southern Florida over the course of the 20th century may have increased the likelihood of hard freezes in the state's warmest reaches [1]. Weather observations over the last few decades suggest a changing pattern of rainfall in regions of the state that varies by the degree of landscape change. In areas in which wetlands have been drained or forests have been cleared to make way for extensive agriculture and urban growth, the frequency of rainfall has measurably declined over time [4]. Because expansive wetlands provide not only the moisture needed for evaporation and rainfall but are also an important source of heat retention during the coolest nighttime hours, the loss of Florida's wetlands could also influence the frequency of subfreezing temperatures across the region.

The most direct evidence for such a landscape-change effect on southern Florida's climate would be provided by a long-term temperature record from the precise locations in which wetlands have been drained and filled over time. However, because these areas were only sparsely inhabited before implementation of the Army Corps' water-management projects, no such temperature record exists. To address this limitation of the available data record, the Colorado State University researchers developed a climate model capable of simulating climate throughout Florida's peninsula with landscape conditions derived from two sets of maps: a pre-1900s landscape compiled from 19th-century surveys of the state and a 1993 landscape derived from satellite imagery (Figure 2.1). By feeding these two sets of landscape conditions into a regional climate model for separate runs and matching all other model inputs to the specific conditions of the January 1997 freezing event, the researchers could isolate the influence of landscape change on this particularly destructive cold snap.

When run with the modern-day land cover, the climate model found temperatures corresponding to the freezing event of January 1997 to be several degrees colder than a model run that included the predevelopment landscape. In some of the hardest hit agricultural areas to the south and east of Lake Okeechobee, the removal of natural marshes and other wetlands resulted in temperatures that were about 3 to 5°F lower than the predevelopment simulation and that increased the duration of subfreezing conditions by several hours. The model results showed that the natural wetlands moderated falling temperatures by releasing heat energy absorbed during the day throughout the night, when temperatures reach their minimum. In these areas, the landscape changes required to facilitate the relocation of orange groves to southern Florida had intensified the very crop losses the relocation was intended to minimize.

The physical principles governing the influence of moisture on climate in southern Florida are the same as those driving climate in other regions endowed with large quantities of water. For example, many a first-time visitor to San Francisco, California, can attest to the surprising chill that can accompany an evening stroll through the city even at the height of summer, whereas others are drawn to the city to experience

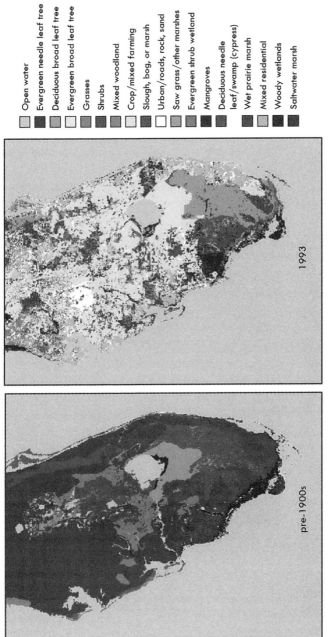

Figure 2.1 Pre- and post-development landscapes in southern Florida. *Source*: C. Marshall, R. Pielke & L. Steyaert, Crop freezes and land use change in Florida, *Nature*, 426, 29–30, 2003. (*A color version of this figure can be found in the color plate section after page 86.*)

Legend:
- Open water
- Evergreen needle leaf tree
- Deciduous broad leaf tree
- Evergreen broad leaf tree
- Grasses
- Shrubs
- Mixed woodland
- Crop/mixed farming
- Slough, bog, or marsh
- Urban/roads, rock, sand
- Saw grass/other marshes
- Evergreen shrub wetland
- Mangroves
- Deciduous needle leaf/swamp (cypress)
- Wet prairie marsh
- Mixed residential
- Woody wetlands
- Saltwater marsh

1993

pre-1900s

winter temperatures that are far more moderate than those found far-ther inland. In each instance, it is the proximity to two large bodies of water, the San Francisco Bay and the Pacific Ocean, that moderates the climate extremes of summer and winter. By the same physical principles governing moisture and weather in San Francisco, the removal of mois-ture in the form of wetlands from southern Florida should be expected to intensify climate extremes – and this is precisely what the Colorado State University researchers found through their climate simulations. Not only were winter dates found to be colder relative to a predevelop-ment landscape, but also summer dates were found to be as much as 7°F hotter than would have occurred in the 19th century. Likewise, sum-mer rainfall was reduced by 10% to 12% after the removal of wetlands and forests, producing a region that was hotter, drier, and generally less suitable to the purposes for which it was modified [4].

Although the enhanced potential for hard freezes in southern Florida would come as an unwelcome discovery to the citrus industry, it should not have. Orange growers have long recognized the benefits of surface water for moderating climate extremes – as shown by the cost premium attached to grove sites located immediately downwind of lakes and ponds and by the reduced value of sites with surface depressions, in which cold air is likely to pool. The climatic benefit of such a "lake effect" is precisely the reason why much of south Florida's orange-growing pro-duction today is clustered immediately to the south of Lake Okeechobee, where the orange groves are warmed in the winter by a southeasterly wind across the lake.

The travails of Florida's orange growers over the past quarter-century are only one manifestation of a larger climate narrative link-ing landscape changes to climate. It is estimated that human actions have modified between one-third and one-half of the planet's land sur-face, most often in the form of extensive deforestation [5]. Although the resulting landscape changes entail profound implications for cli-mate at local and regional scales – the scales at which humans experi-ence extreme temperatures and storm events – these climatic effects are largely unaccounted for in the IPCC's global-scale climate models. Much like the Florida orange growers who assumed that the strong linkages between landscape and climate observed at the scale of the grove were not in effect at the scale of the region, the climate-science community has largely ignored mechanisms of climate change that are assumed to be inoperable at the global scale. As a result of this oversight, the most commonly advocated strategies to mitigate global warming will not prove sufficient to safeguard human health in many regions of the planet.

To observe that wetlands destruction, deforestation, and other land surface changes are important drivers of climate change is not to diminish the significance of greenhouse gases. Rather, it is to assert that the greenhouse effect is not driven by greenhouse gases alone. To understand why this is the case, it is useful to revisit the basic physical workings of the global greenhouse effect.

As theorized by Fourier and later demonstrated by Tyndall, greenhouse gases such as carbon dioxide and methane warm the atmosphere by trapping outgoing terrestrial or longwave radiation emitted from the Earth's surface. The key physical property of greenhouse gases governing this warming effect is their selective nature with respect to the absorption of radiation: greenhouse gases do not absorb incoming, shortwave radiation from the Sun – largely consisting of visible light – but do absorb outgoing, longwave radiation from the Earth's surface – largely consisting of invisible thermal-infrared radiation – thereby serving to warm the atmosphere. Rising concentrations of greenhouse gases in the atmosphere are enhancing the global greenhouse effect by increasing the quantity of outgoing, terrestrial radiation that is trapped by the atmosphere.

It is well established that the concentration of greenhouse gases present in the atmosphere regulates the quantity of outgoing terrestrial radiation absorbed by the atmosphere. Yet, what physical properties govern the quantity of terrestrial radiation emitted from the Earth's surface in the first place? It is with respect to this component of the greenhouse effect that the physical properties of the landscape play a central role in climate change.

Although numerous properties of the Earth's surface influence the quantity of longwave radiation emitted to the atmosphere, two merit discussion here. The first property is the reflectivity of the Earth's surface or, as it is described more formally, its "albedo." Albedo quantifies the proportion of shortwave radiation received from the Sun that is reflected back out to space. Importantly, the mechanism of reflection entails no change in the nature of the radiation itself. Were the surface of the Earth a perfect reflector such as a mirror (i.e., characterized by an albedo of 1 or 100%), all of the shortwave radiation received from the Sun would be reflected back to the atmosphere in the form of shortwave radiation. Were this to occur, the reflected shortwave radiation mostly would pass back through the atmosphere and out to space because greenhouse gases generally do not absorb shortwave radiation, rendering the planet a much colder place.

When sunlight is intercepted by surfaces with less than 100% albedo, a category that encompasses all of the Earth's natural land

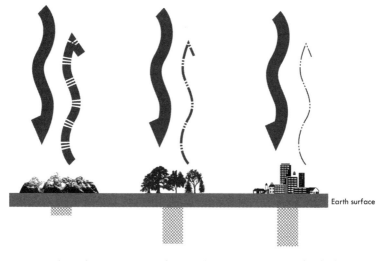

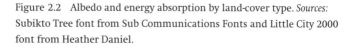

━━━━ solar radiation ━━ıı━━ reflected solar radiation ⬚⬚⬚⬚ absorbed energy

Figure 2.2   Albedo and energy absorption by land-cover type. *Sources:*
Subikto Tree font from Sub Communications Fonts and Little City 2000
font from Heather Daniel.

covers, some percentage of the intercepted shortwave radiation is
absorbed rather than reflected back out to space (Figure 2.2). Once
absorbed, some percentage of this energy will be returned to the atmosphere in the form of longwave radiation. Thus, the degree to which the
greenhouse effect warms the planet is driven, in part, by the proportion
of intercepted solar radiation that is absorbed at the Earth's surface and
converted into longwave radiation. Albedo governs the quantity of longwave radiation emitted to the atmosphere by limiting the amount of
solar radiation that is absorbed in the first place. Every time we humans
fill a wetland, clear-cut a forest, or construct a road, we alter the albedo
of the Earth's surface, sometimes in ways that increase the quantity of
longwave radiation emitted to the atmosphere.

A second key physical property governing the production of heat
energy is the "energy balance" of the Earth's surface. The energy balance
refers to the specific mechanisms through which absorbed solar energy
is returned to the atmosphere. The absorption of solar energy at the
surface of the Earth will increase the temperature of surface materials in
the path of sunlight. At some point, these surface materials will reach a
threshold temperature at which the temperature of the material, such as
the sand on a beach, is higher than the temperature of the overlying air.

At this point, increasing quantities of heat will be released from the land to the atmosphere to approach an equilibrium in temperature between a surface material and the surrounding environment. Heat energy can be returned to the atmosphere from the land surface through one of three basic mechanisms.

The first mechanism, as noted earlier, is the re-emission of radiation in the form of longwave radiation. Such thermal-infrared radiation, characterized by a longer wavelength than the visible light of the Sun, is the predominant form of heat that we experience in warming our hands by a campfire. Because of this outgoing terrestrial radiation's longer wavelength, a large percentage of it is absorbed by greenhouse gases, which contributes to an increase in atmospheric temperatures through the global greenhouse effect (see Figure 1.2).

A second pathway or "flux" through which heat energy is returned to the atmosphere is through the heating of the air immediately above the Earth's surface. As the temperature of sand on a beach increases, the air immediately above will also warm and then rise, distributing heat through the atmosphere via the process of convection. This form of heat energy is known as "sensible" heat because it can be detected directly as a rise in the temperature of the air. The heat we experience upon opening an unventilated car in summer is a familiar example of sensible heat.

A final flux through which heat energy is returned to the atmosphere is in the form of "latent" heat. Latent heat is so named because its transmission does not result in a detectable rise in temperature. Rather than contributing to an increase in the temperature of a receiving surface, some proportion of the intercepted solar radiation may be used to evaporate water found within the surface material, such as the moisture present within wet sand. "Transpiration" is a similar process through which water vapor is released from the leaves of plants. As a byproduct of bringing about a phase change in water – the conversion from water to water vapor – the intercepted heat energy remains locked up in the water vapor that is released to the atmosphere. Now in a latent, nondetectable form, this energy is transported by the rising water vapor to the upper atmosphere, where the water vapor will eventually cool, condense back into water droplets, and contribute to cloud formation. Each of these heat-transfer mechanisms and their relative magnitudes in urban and rural environments is illustrated in Figure 2.3.

The significance of the latent heat flux is that it permits solar radiation absorbed at the Earth's surface to be returned to the upper atmosphere without contributing to a rise in temperature at the surface. Thus, land features that increase the availability of moisture for

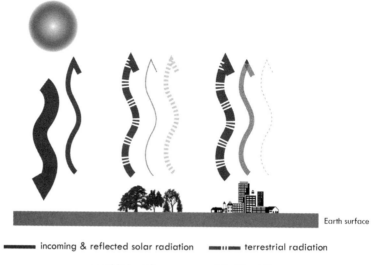

Figure 2.3   Mechanisms of heat transfer from the land surface to the atmosphere. *Sources:* Subikto Tree font from Sub Communications Fonts and Little City 2000 font from Heather Daniel.

evaporation and transpiration (referred to in shorthand as "evapotranspiration"), such as wetlands or forests, can reduce the temperature of the Earth's surface during warm periods and, in so doing, reduce the quantity of longwave radiation and sensible heat emitted to the atmosphere. The extent to which heat can be transferred in a latent form depends in large part on the availability of moisture at the surface. Wet sand, for example, exhibits a lower temperature than dry sand even with the same receipt of solar radiation because of its ability to cool itself through evaporation.

The surface energy balance can thus be understood as a terrestrial valve that controls the flow of solar energy back to the atmosphere. Landscapes endowed with ample quantities of moisture will return more of that energy to the atmosphere in a latent form that contributes little to enhanced temperatures at the surface, whereas landscapes lacking moisture will return more of that energy back to the atmosphere as longwave radiation and heated air.

In combination, the surface properties of albedo and energy balance play a significant role in the warming of the atmosphere. If human modifications to the landscape are serving to reduce albedo and the availability of moisture for evaporative cooling, a larger quantity of intercepted solar energy will be returned to the atmosphere in the form of longwave radiation, driving temperatures higher even without an increase in the atmospheric concentration of greenhouse gases. In addition, such changes will bring about higher temperatures at the Earth's surface through the increased emission of sensible heat.

The global greenhouse effect, therefore, is not a product of greenhouse gases alone but instead is driven by two related but independent mechanisms. In the most basic terms, these two warming mechanisms can be understood to function much like a pot of boiling water. On the one hand, the temperature of the water can be raised by sliding a lid over the top of the pot. As the lid traps outgoing steam, the temperature of the water will rise, independent of any increase in the intensity of the burner's flame. This is analogous to the effect of rising concentrations of greenhouse gases. On the other hand, the temperature of the water can be elevated through increasing the heat source underneath, with no change in the position of the pot's lid. This is analogous to the effect of land surface changes that reduce surface albedo or the latent heat flux. Although both actions have direct and significant effects on temperature, the global climate community to date has tended to focus on the role of the lid rather than that of the burner. This limited focus comes at a cost.

The implications of land-surface changes for climate are perhaps nowhere more apparent than in the Amazon basin of South America. A moist broadleaf rainforest about three-quarters the size of the continental United States, the Amazon has been subject to accelerating rates of deforestation since the post–World War II era, when Brazil, Peru, and other countries into which the vast forest expands started constructing roads into its interior. Before that time, the rainforest was almost impenetrable, with the most remote areas requiring a minimum of six weeks to reach from the Brazilian metropolises of Sao Paulo and Rio de Janeiro by way of the Amazon River. The roads were intended to open up the ancient forest to development and resource extraction, and they have served these purposes well. Yet the benefits of land clearance in the Amazon have often proved to be short-lived.

The largest tropical rainforest in the world, the Amazon basin receives prodigious amounts of rainfall over the course of a year, averaging in some regions more than an inch a day. In combination with year-round warm, equatorial temperatures, the extreme moisture levels of the rainforest give rise to a very rapid recycling of soil nutrients back into plant and animal life. One byproduct of this rapid recycling of nutrients is that the biologically active layer of soil in a rainforest environment is very shallow and must be continually replenished through decomposing organic matter to sustain the dense vegetation. Absent this replenishment, the soils of much of the Amazon basin are poorly suited for intensive cultivation. As a consequence, the "slash and burn" agriculture practiced in the rainforest, through which peasant farmers clear land by burning and then plant with crops, often requires new land to be cleared every few years, once the nutrient-poor soils of the denuded rainforest land have been depleted.

In recognition of the inherent limitations of cleared rainforest land for agriculture, the Brazilian government has provided generous subsidies since the 1960s to attract investment in the Amazon basin. Policies such as allowing corporations to write off up to 50% of their annual tax liability by investing in rainforest development projects, or providing tax credits for up to 75% of the capital costs of new development projects, have greatly fueled corporate investment in the basin, and much of this investment has resulted in the clearing of rainforest land for cattle ranching. Responsive to Brazil's insatiable demand for beef and entailing low labor costs, cattle ranching is today a highly profitable industry in the Amazon. Although the soils of the cleared land are only marginally more suitable for pastureland than for crops, ranchers have compensated for this limitation by simply clear-cutting larger swaths of the forest to support cattle. It is estimated that almost 6 acres (2.4 hectares) of land are required to support a single head of cattle in these regions, among the lowest cattle densities in the world [6].

In combination, the aggressive clearance of land for small-scale farming and large-scale cattle ranching has resulted during the last half-century in a loss of rainforest area equivalent in size to the northeastern United States. In addition, the rate of deforestation has accelerated in recent years, with 18% more land cleared between 2000 and 2005 than during the preceding five years [7]. Although the loss of rainforest has an immeasurable impact on global biodiversity – one in 10 of all known species live in the Amazon – deforestation of tropical rainforest also entails significant implications for climate change.

It is estimated that about half of the rain that falls in the Amazon basin is the product of evapotranspiration and cloud formation within the basin itself. As rainforest is cleared to make way for nonirrigated pastureland, the quantity of moisture retained by the soils and plant life of the cleared area is only a fraction of that of the native rainforest landscape, thus reducing the availability of moisture for evaporative cooling. As less solar energy is used in the evaporation of water, more is used to increase the temperature of the surface and near-surface air, resulting in higher temperatures. And, as less moisture is evaporated, less water vapor is available in the basin for cloud formation and subsequent rainfall.

The degree to which deforestation in the Amazonian basin is chang- ing regional climates has been the focus of numerous studies since the 1980s. A number of observational studies have recorded reductions in cloud formation and rainfall, and subsequent increases in temperature, following the removal of tropical rainforest land covers. Because these studies tend to focus on land-surface changes over relatively small geo- graphic areas (e.g., land cleared for a single farm), regional climate models have often been used to assess the implications of basin-wide land-cover changes for climate. The results of these studies suggest that land-use changes can be a powerful driver of warming.

In light of the rapid clear-cutting of rainforest to make way for pastureland, several studies have estimated the impacts of a complete degradation of the Amazon basin to grasslands. As theory and local-scale observations would suggest, these studies find the loss of rainforest to result in significant reductions in evapotranspiration, rainfall, and cloud formation, accompanied by increases in average basin-wide tempera- tures of about 2 to 7°F [8–10]. Comparable in magnitude to the warming observed in southern Florida after wetlands removal, this rise in temper- ature is roughly equivalent to the midpoint of projected global warming by the end of the century.

In concert with landscape changes, rising concentrations of green- house gases will also increase temperatures in the Amazon basin over time. What is the relative contribution of both land-surface changes and greenhouse gas emissions to regional climate change? To address this question, climate models have been used to account for a transition from a rainforest habitat to grasslands or desert in combination with projected increases in atmospheric concentrations of greenhouse gases. The results show that land-surface changes are responsible for more than 40% of the modeled warming in the Amazon resulting from human changes to the local and global environments [10].

These studies highlight a rather surprising conclusion with respect to climate change in at least one region of the planet: the actual impacts of human activities on climate at the regional scale, accounting for both land-surface changes and emissions of greenhouse gases, may be *twice as great* as the impacts of greenhouse gases alone. What is most worri- some about these findings is that the suite of climate models on which IPCC projections of future warming are based often do not account for the direct influence of land-surface changes on climate [11, 12]. Thus,

Figure 2.4 Rainforest land clearance in the Amazon basin viewed from space. *Source:* NASA's Earth Observatory, http://earthobservatory.nasa. gov/Features/WorldOfChange/deforestation.php.

although these models might account for the impact of Amazon defor-estation on CO_2 emissions resulting from the burning of rainforest, they typically ignore the substantial increase in regional temperatures result-ing from changes to the surface energy balance. More worrisome still is that the influence of ongoing land-surface changes on climate is not lim-ited to the Amazon basin but instead is found in every part of the planet.

One reason that global climate-change modeling often overlooks the impacts of deforestation and other land-use changes is that they are typi-cally carried out in a spatially heterogeneous pattern over long periods of time. In contrast to the globally uniform rise in CO_2 concentrations, land-use changes are highly irregular over space and thus are more difficult to associate with a consistent climate trend. As illustrated in Figure 2.4, rainforest destruction in the Amazon adheres to what has been described as a "herringbone" pattern: linear incursions into the jungle interspersed with remaining fragments of forest canopy. One result of this irregular pattern of land-use change is that no clear boundary exists between altered and natural landscapes that might give rise to a sharper division in climatic conditions apparent to anyone crossing the boundary. Rather, observations and computer modeling only reveal the cumulative effects on climate of small-scale land-use change over long periods of time.

There is one place in the world, however, where there is a sharp division between natural and human-altered landscapes over an expansive geographic area. Known formally as the State Barrier Fence, a 730-mile fence across Southwest Australia partitions a large agricultural zone to the west from native forest and scrublands to the east. Its history is rooted in the importation of rabbits to Australia for hunting by English settlers in the 19th century, which, as with most nonnative species introductions, resulted in a range of unintended consequences. Within only a few years of the arrival of rabbits in the Outback, settlers were boasting of their ability to shoot as many as 1,200 of the prolific breeders in a single afternoon. Yet, the hunting parties would be short-lived because the settlers who made their living as farmers came to grips with the destructive potential of an exploding population of herbivores with no natural predators. Less than 30 years after the introduction of rabbits to the Outback, the Intercolonial Royal Commission was offering a £25,000 prize to anyone who could devise a solution to the "rabbit menace" [13].

It was believed then that the solution was to be found in a vermin fence, and so, in 1901, the Australian government would undertake the massive task of constructing and maintaining a rabbit-proof fence across a large expanse of the Australian continent. Although the fence has never proven particularly effective in controlling the spread of rabbits, it has provided a clear demarcation between lands perceived to be suitable for agriculture and those that are not, and its effectiveness in transforming the native vegetation of Southwest Australia is indisputable. As illustrated in Figure 2.5, the State Barrier Fence has created a sharp boundary between the once-dominant native scrublands to the east and the expansive croplands of wheat and barley to the west. It has also created a well-designed natural experiment to assess the influence of land use on climate change.

The first clear evidence that the widespread changes in land cover to the west of the fence might have climatic implications appeared in the 1970s. The onset of a severe drought by the middle of that decade would signal the arrival of fundamental change in the region's climate; the drought was in fact not a temporary shortfall in regional precipitation but rather marked the sudden onset of a long-term shift in rainfall patterns from the west to the east. Since the mid-1970s, areas to the west of the barrier fence have experienced a 15% to 20% reduction in rainfall during the growing season, with serious consequences for population centers dependent on these watersheds for drinking water. The City of Perth, for example, with a population of 1.7 million, has experienced a reduction in inflows to regional water supplies of more than 40%, a dramatic

Figure 2.5   Diverging land-use patterns along the State Barrier Fence in Southwest Australia. *Source:* NASA, Goddard Space Flight Center.

shortage in drinking water that has required the rapid and costly construction of a new network of reservoirs. As to be expected, the drier conditions to the west of the fence have been associated with rising temperatures there as well [14].

Surmising that the land-use changes to the west of the State Barrier Fence have played a role in bringing about rapid changes in regional climate, a handful of scientists have sought to quantify the relative contributions of changing land use and greenhouse gas concentrations to the observed shift in regional climates. With the aid of satellite imagery and land-based measurements, scientists have shown that changes to the native vegetation have resulted in substantial changes to albedo and the surface energy balance to the west of the fence, with direct implications for the water cycle. Although the cropland to the west is found to have a higher albedo throughout the year, the cooling benefits of this higher reflectivity are more than offset by a substantial reduction in moisture availability during the warm season, when temperatures are highest. As a result of lower soil-moisture availability in the summer, evaporative cooling has been found to be an average of 60% lower on the western side of the barrier, requiring a larger percentage of absorbed solar energy to be returned to the atmosphere as sensible heat [14, 15].

The implications of reduced evapotranspiration for weather in Southwest Australia have been significant. Cloud formation has been shown to be almost 70% lower to the west of the barrier fence during the summer than to the east, over land with native vegetation. The direct influence of land-surface conditions is illustrated well in Figure 2.6, which clearly shows the prevalence of cloud formation immediately to the east of the State Barrier Fence. In this instance, the native vegetation acts as a reservoir for moisture and makes use of this moisture to offset surface-heat gain through evapotranspiration. This evapotranspiration drives cloud formation and, ultimately, rainfall, resulting in conditions that are wetter to the east of the fence than to the west in the summer season, when the winter croplands for wheat and barley remain bare. As a result of these changes, average summer temperatures have been shown to be several degrees warmer in the regions converted from natural vegetation to croplands [15].

Changing land-use conditions not only create regional differences in climate on either side of the barrier fence but also contribute to warming trends over time. Similar to research in the Amazon, climate models have been employed in Southwest Australia to estimate the degree to which land-use changes are driving regional warming trends relative to rising levels of greenhouse gases. Much like in the Amazon, these modeling studies find that the displacement of natural forest and shrublands by agriculture has contributed to as much as half of the observed warming since the early 20th century, with the global greenhouse effect accounting for the remainder [14].

In Southwest Australia, human alterations of the landscape are a major driver of climate change. And although the clear scientific evidence supporting this conclusion suggests that the problem of climate change is more complex than the problem of greenhouse gas emissions alone, born of this complexity is the potential for Australians to exert far more control over their own climate future than is possible through emissions reductions alone. For, whereas a complete cessation of greenhouse gas emissions in Australia today would have little impact on the pace of climate change in the near term, a partial restoration of natural

Figure 2.6   Preferential cloud formation to the northeast of the State
Barrier Fence in Southwest Australia. *Source:* D. Ray, U. Nair, R. Welch, et
al., Effects of land use in Southwest Australia: Observations of cumulus
cloudiness and energy fluxes, *Journal of Geophysical Research*, 108, 4414,
2003. Reproduced/modified by permission of American Geophysical
Union.

land covers west of the State Barrier Fence could enhance regional rainfall and moderate the pace of warming within the span of a decade
or less. Because both the global greenhouse effect and regional landscape alterations are driving climate change in Southwest Australia, both
drivers should be the subject of climate management policy. Yet today
in Australia, as throughout the world, the problem of climate change is
generally understood to be a problem of greenhouse gases alone.

The significance of forested landscapes to climate observed in the Amazon and Southwest Australia holds true for most regions of the planet.
Much like the lost wetlands of south Florida, forests tend to act as

reservoirs of moisture, moderating climate extremes by retaining heat in the winter and offsetting additional heat gain through evapotranspiration in the summer. As documented by numerous studies on forest cover and climate, tropical to mid-latitude forests play a significant role in slowing the pace of climate change due to both the sequestration of carbon dioxide and evaporative cooling. It is only in the most poleward reaches of the planet, where the presence of snow cover throughout much of the year creates a very high surface albedo, that low-albedo conifer forests are associated with a greater degree of warming than cooling [16]. Yet, even in these environments, forest cover is accomplishing a consistent ecological objective: rendering an extreme climate more moderate.

In the United States, the influence of forests and other natural land covers on climate has been measured in several ways. Studies of abandoned agricultural plots that have been permitted to revert to forest find the reforested land cover to be associated with a reduction in near-surface temperatures of about 4°F [17]. Likewise, larger scale climate modeling studies, in which the effects of reforestation across large regions of the eastern United States are simulated, find a transition from croplands to forest to be associated with a widespread cooling effect in summer months [18, 19].[1]

Perhaps the most compelling evidence of the role of land use in climate change is provided by a comparison of upper atmosphere, weather-balloon measurements of temperature with surface-based measurements from weather stations. Because high altitude weather-balloon data is insensitive to the influence of land use on climate, such as the displacement of forests by croplands or cities, these measurements can be subtracted from land-based measurements of temperature at the same location to isolate the influence of land use on climate trends. Through this approach, scientists have estimated that deforestation and other land-use changes have resulted in about 0.2°F of warming per decade across the eastern United States since the 1960s – a rate of warming that

---

[1] It should be noted that some global-scale modeling studies find evidence of a warming effect produced through the conversion of croplands to forest in the United States. Whereas this discrepancy may result from the differing scales at which climate simulations are run (i.e., global versus regional scales), important model assumptions, such as the extent to which croplands are actively irrigated, may also contribute to divergent results. No such discrepancies characterize modeling or observational studies of tropical deforestation, which is consistently found to drive surface warming and accounts for most global deforestation. See [16] for a detailed discussion of the climatic effects of forests at different latitudes.

accounts for more than half of the observed temperature changes during this period (with the global greenhouse effect believed to account for the remainder) [20, 21]. Were this rate of warming to continue throughout the current century, deforestation and other land-use changes would result in a temperature increase of about 2°F by 2100, independent of ongoing changes in levels of greenhouse gases.

From what is today an extensive body of scientific work, focused on the United States and elsewhere, a general rule of thumb is emerging: *Deforestation and other alterations of the Earth's land surface account for as much as half of the warming trends underway in the regions in which these changes are occurring.* And a critically important corollary to this rule of thumb is: *Global-scale climate change projections that fail to account for the regional effects of land-use change are likely to significantly underestimate the true extent of future warming at the regional scale.*

If true, however, why then does the global climate-science community in general, and the IPCC in particular, fail to fully account for future land-use changes into global-scale climate projections? The most direct answer to this question is that most global climate scientists presume these effects to be small. Indeed, when measured at the global scale, the climatic effects of land-use change do appear to be small. However, this outcome is simply an artifact of the adoption of the global scale as the appropriate dimension to measure climate change.

It is useful to consider the reasons why regional land-use changes have only a limited influence on climate when measured at the global scale. First, the global land mass accounts for only about one-third of the Earth's total surface area, with the oceans accounting for the remainder. Although changes to the land surface entail direct implications for the temperature and moisture of the atmosphere in proximity to land, these land-surface changes may have relatively little effect on the two-thirds of the planet's surface covered by water. Thus, the averaging of land-use effects across the entire surface area of the Earth statistically diminishes the magnitude of these effects in the places that they are actually experienced by people – on land. By contrast, as a globally diffuse climate-change agent, greenhouse gases warm the atmosphere over both land and ocean, thus producing a warming effect that is appropriately measured on a global scale.

A second reason that the climatic influence of land-use change is difficult to detect at the global scale pertains to the offsetting of these effects by season and between regions. As demonstrated by the loss of wetlands in Florida, human alterations to landscapes often diminish regional moisture availability. As a result, these alterations may enhance climate extremes by season, with summers growing hotter and winters growing somewhat colder as well. When measured as a change in annual average temperature, these seasonal effects tend to offset one another, potentially resulting in only a marginal change in average temperatures. Here again, the statistical averaging of climate data substantially alters its significance.

The observation that average annual temperatures have not increased much in your city, despite the fact that summers have become substantially hotter, would likely be of little consolation. Yet this is the way in which climate change is generally measured. Likewise, when averaged at the global scale, land-use changes that produce a net warming effect in one part of the planet may be offset by land-use changes creating a net cooling effect in another. Here again, even highly pronounced impacts at the regional scale are often statistically obscured by the derivation of global averages.

Finally, the poor resolution of global climate models generally does not permit the accurate measurement of regional-scale land-use changes. Global climate models partition the Earth and atmosphere into a large number of three-dimensional boxes or grid cells, each covering a portion of the Earth's surface or the atmosphere above the surface. Due to the large number of physical and chemical equations that must be solved for each cell, the total number of cells for which climate can be modeled is limited by computer-processing speed. In the interest of minimizing processing times, the size of grid cells can be quite large, often covering a surface area of 5,000 square miles or more – larger than the area of some U.S. states. At this large dimension, land-surface changes impacting only a fraction of the cell may have only limited effects on the cell's average climate conditions. Thus, the clear-cutting of an area of forest the size of a large U.S. city may not be sufficient to significantly change average climate conditions across a much larger grid cell. As a result, most regional-scale land-use changes simply cannot be seen by these global-scale models.

Therefore, in the most basic terms, global climate scientists tend to overlook the climate effects of deforestation and other land-use changes because global-scale climate models do not show these effects to be significant. But this observation suggests a more fundamental question: Why is it useful to model climate change at the global scale? It is certainly true that many of the climatic conditions experienced at any regional location are influenced by global or hemispheric phenomena, such as jet streams and oceanic circulations. Yet, weather events, the day-to-day expression of climate, do not occur at global or hemispheric scales. Severe storms,

floods, heat waves, droughts, and, perhaps most important, daily fluctuations in temperature only occur in any experiential way at the scale of
regions. Thus, although it is possible to statistically describe climate at
the level of the planet as a whole, no human being experiences a globally
averaged climate. Global average climate, in this sense, is nothing more
than a statistical abstraction.

Despite this fact, mean global temperature change has been universally adopted as the metric by which climate change is measured.
Climate scientists have developed extensive networks of land-, sea-, and
space-based monitors for the purpose of measuring changes in mean
global temperatures, and annual updates of mean global temperature
records are perhaps the most widely recognized evidence of climate
change. Based on the unquestioning acceptance of this metric by policy makers, the suitability of climate-management strategies is directly
assessed against their potential to influence mean global temperatures,
even if those strategies are unlikely to soon slow warming in the locations where most of the human population resides. Given the emphasis
on global measures, many nations are unwilling to take individual action
on climate change absent a binding international agreement on strategies designed to reduce mean global temperatures.

The evidence reviewed in this chapter demonstrates clearly that to
understand climate change in global terms alone is to fail to understand
its full complexity. The climate experienced at any location on the planet
is strongly influenced by both global and regional mechanisms, and both
mechanisms offer the potential to slow warming trends. To attempt to
manage climate change through a global emissions reductions strategy
alone is to overlook entirely a set of human activities that account for
as much as half of the observed warming trend in many regions of the
planet. Yet, this today is the international status quo in the realm of climate change policy. Developing a more comprehensive and, ultimately,
effective approach to climate-change management will require a broader
set of perspectives on the underlying drivers of warming than that provided by the atmospheric sciences alone, and it will require policy tools
that enable governments to move forward in managing climate-related
challenges absent a politically elusive international consensus. The uni-
versal adoption of globally averaged metrics impedes both objectives.

Nowhere is the adherence to global indicators of climate change
more problematic than within large cities. As explored in the next chapter, many cities have already exceeded the magnitude of warming projected for the planet as a whole over the current century. Driven by the
same physical principles at work in the Amazon or Southwest Australia,

but on a far more profound scale, the dramatic warming that is occurring in the most heavily populated regions of the planet is not revealed through global measures of climate change. To the contrary, climate scientists have developed techniques to statistically remove the enhanced warming associated with cities from weather-station data obtained in urban areas. Unaware of these trends, relatively few scientists or government officials are attempting to track the rate of warming occurring in cities or to assess its implications for extreme heat events and human health. For many, to live in a large city today is to live on the leading edge of the most rapidly changing environmental conditions ever experienced by humans – and to not even know it.

3

Islands of Heat

At 4:10 p.m. on August 14, 2003, engineers in the control room for the New York Independent System Operator (NYISO), the organization charged with operating the state's electrical transmission grid, were alerted to a sudden and massive surge of power to the western region of the state. Within seconds, the surge had increased in intensity and pulsed back into the New York system, activating a series of automated breakers that would sever the state's grid into two divisions in an attempt to forestall a complete shutdown. Gaping at the NYISO's wall-sized control screen, system operators watched powerlessly as the hundreds of lights representing core components of the state's electrical grid turned red in sequence and started flashing. By 4:14 p.m., the economic capital of the world had disappeared from computer screens across the planet, and no one knew why.

The NYISO operators were witnessing in real time the beginnings of the most massive power outage in American history. Within a few minutes on an otherwise unremarkable day, 55 million residents of Midwestern and Northeastern cities, as well as portions of Canada, had lost all power. Assuming the interruption to be localized and likely of brief duration, many New Yorkers sat quietly in front of blank computer screens in their darkened offices, while others tried to remain calm on stranded subway cars under the city's streets and in the arrested elevator shafts of the city's buildings. As the magnitude of the event soon became apparent, workers started filing out of buildings onto the city's hot streets to make their way home – on foot (Figure 3.1).

As the rest of the country watched images of civilians directing traffic underneath darkened traffic signals, while thousands streamed on foot over the Brooklyn Bridge – with others still left to spend the night on the sidewalks of Manhattan – the 2003 blackout would leave a clear and unsettling impression: the technological veneer partitioning

Figure 3.1   Stranded New Yorkers cross Brooklyn Bridge on foot during a
multi-state blackout on August 14, 2003. *Source:* Jonathan Fickies / Getty
Images News / Getty Images, 2003.

the modern from the premodern worlds had become uncomfortably
thin.

The ultimate cause of the blackout was the heat. Although less intense
than the heat wave underway across the Atlantic that summer, a short-
term heat spell in the northeastern United States was spiking the regional
demand for electricity. Because of a failure to sufficiently invest in transmission infrastructure for decades, many regions of the United States rely
on a relatively small number of transmission corridors to carry excessive
electrical loads during periods of high demand. In one such corridor
in rural Ohio, the demand grew so intense that the high-voltage power
lines started to physically sag in response to the intense heat associated
with extreme electrical loads. Several feet lower than their normal operating heights, the overstressed power lines came in contact with tree
limbs underneath and short-circuited, disabling the transmission line.
The sudden drop in electrical transmission to nearby cities triggered a
surge in power from the east to compensate, creating in turn a power
void in the New York region and requiring a surge from suppliers to
the west. To avoid damage to generating plants caused by the cascading
power surges, the regional system soon shut down entirely.

As the heat subsided over the next 24 hours, power-generating plants were slowly brought back online and the regional grid was restored. For the most part, New Yorkers and other residents paralyzed by the massive blackout had weathered the event well, with many taking delight in the brief return of insect noises and discernible constellations to the stilled night sky. Yet, even a brief disruption of power to New York had resulted in loss of life, with at least six residents succumbing to the heat in a single night [1]. In light of this experience, one is left to ponder the implications of an increasingly likely scenario: the concurrence of a prolonged heat wave with a multiday blackout.

The growing potential for such an event raises troubling questions about the preparedness of cities to cope with even a modest continued rise in summer temperatures. For, as New Yorkers experienced firsthand, electricity powers not only our computers, refrigerators, and air conditioning but also, in some fashion, all forms of mechanized transport – from the subways under the street to the fuel pumps at gas stations to the control towers at airports.

More significant even than transport or air conditioning is the delivery of water, which is driven by a vast network of electrical pumps. If a multiday blackout were to occur in response to a prolonged heat wave, what emergency systems would be in place to sustain large urban populations unable to move beyond the blackout zone and lacking access to clean water? What emergency systems of the needed scale would even be possible?

Climate change is increasing the likelihood of such a scenario by increasing the frequency and intensity of heat waves. In theory, as the mean annual temperature of the planet increases, the frequency of heat waves should increase as well, assuming the historical threshold for defining a heat-wave event remains unchanged over time. To test this theory, my research group at the Georgia Institute of Technology, the Urban Climate Lab, collaborated with the U.S. Centers for Disease Control and Prevention to analyze daily weather records for more than 50 of the largest U.S. cities from the 1950s to the present. Consistent with other studies of climate change, we measure temperature trends over a multi-decadal period to account for any shorter term shifts in climate – such as those related to the 11-year solar cycle or the 2- to 7-year El Niño Southern Oscillation – that might give rise to a temporary increase in heat wave occurrence.

In measuring the frequency of heat waves over time, it is important to note that what constitutes an unusually hot day in a region like Phoenix is different from what passes for an unusually hot day in Chicago. The reason for this is due not only to differences in latitude and climate between Phoenix and Chicago but also, more importantly, to the acclimation of each region's population to a "normal" range of temperatures. Acclimated to the hot and dry weather of a Phoenix afternoon, residents of that region are unlikely to experience widespread heat stress from afternoon temperatures in the range of 100°F, whereas emergency rooms in Chicago would be expected to experience a pronounced uptick in heat-related conditions in response to such temperatures.

Therefore, what constitutes an unusually hot day in any particular region is based on its departure from a long-term normal range of temperatures that is specific to the region. Referred to technically as an "extreme heat event," such events are defined as any day in which the temperature exceeds the 85th percentile of the long-term range of temperatures experienced in that region [2]. In my own region of Atlanta, Georgia, for example, any day in which the high temperature exceeds 94°F or the low temperature exceeds 77°F would be characterized as an extreme heat event. The 85th percentile of the long-term distribution of temperatures is used as a threshold because studies have found that hospital admissions for heat-related conditions typically occur in excess of this threshold.

Using this approach, we found the number of extreme heat days per year in large U.S. cities to be increasing rapidly over time. On average, the typical large U.S. city experienced about 10 extreme heat days annually in the mid-1950s. By the mid-2000s, this number had doubled to more than 20 such annual events. This is a remarkable number: for the equivalent of about three weeks a year, temperatures are hot enough in U.S. cities to send city dwellers to the hospital. Yet even more remarkable is the pace of change: the frequency of extreme heat events has increased by more than 100% in just 50 years, a period of time during which the planet's average temperature has increased by little more than one degree [3].

Of course, these numbers are for the average American city. As illustrated in Figure 3.2, many cities are experiencing a much more rapid increase in the occurrence of extreme heat. In the last decade, half of these cities have experienced more than a month of extreme heat days in a single year. In these years, one out of three summer afternoons is dangerously hot, and these temperature extremes are not limited to the country's Sun Belt alone. Salt Lake City, Utah; Columbus, Ohio; and Syracuse, New York, have each experienced more than 30 extreme heat days in a single year in the last decade; San Francisco, known for its

72

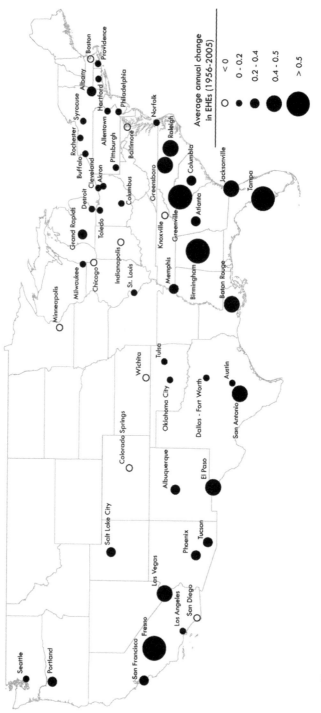

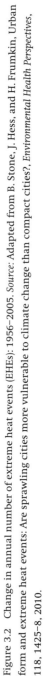

Figure 3.2 Change in annual number of extreme heat events (EHEs): 1956–2005. *Source:* Adapted from B. Stone, J. Hess, and H. Frumkin, Urban form and extreme heat events: Are sprawling cities more vulnerable to climate change than compact cities?, *Environmental Health Perspectives*, 118, 1425–8, 2010.

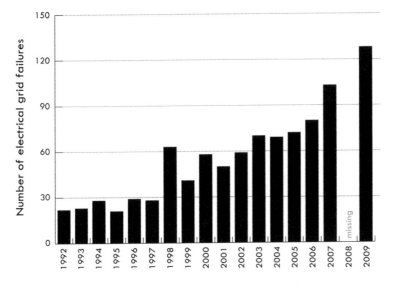

Figure 3.3   Annual electrical grid failures in the United States:
1992–2009. *Source:* North American Electric Reliability Corporation (data
for 2008 are unavailable).

cool summer evenings, experienced more than 50 extreme heat events
in 2003 and more than 70 in 2004.

With such pronounced increases in extreme weather comes not
only an increased likelihood of heat-related illnesses but also increased
stress on the electrical systems powering these cities. Today, the
widespread use of mechanical air-conditioning systems has safeguarded
much of the U.S. population from the most threatening impacts of excessive heat. However, as illustrated by the 2003 blackout, even highly localized failures in the power-generation system can bring down vast regions
of the nation's grid. Recent trends in the frequency of blackouts and other
electrical grid disruptions show a pronounced increase during the past
15 years. As illustrated in Figure 3.3, from 1993 to 1997, there was an average of 26 grid disruptions per year in the United States. Only a decade
later, during the period from 2003 to 2007, that number had more than
tripled to an average of 79 disruptions per year. In combination with
the marked growth in the frequency of heat waves, this is a problematic
trend: the number of grid failures in the United States is increasing at
an average rate of 16 percent a year.

Although all U.S. cities are increasingly vulnerable to increased
heat-wave frequency and potential for blackouts, our study found the
number of extremely hot days in the most sprawling U.S. cities to be

growing at more than twice the rate of the most compact cities. For example, more recently developing and fast-growing cities such as Atlanta, Greensboro, and Phoenix were found to be more prone to extreme heat events over time than older, slower-growing cities such as Boston, Baltimore, and Chicago. The reason for this difference is not due to the Sun Belt location of the former cities but rather to the rate at which their natural landscape is being modified. Much like recent trends in the Amazon basin and Southwest Australia, the rate at which natural land cover is being lost in these cities is an equivalent or more powerful driver of climate change than the emissions of greenhouse gases. Yet, the climate-related impacts of deforestation in cities are far greater than those of deforestation within rural areas. Known formally as the *urban heat island effect*, landscape change in cities is emerging to be the principal climate-related threat to human health today, yet this phenomenon receives relatively little attention within the climate-science community. This oversight is proving to be costly.

The emergence during the early to mid-20th century of a "new towns" planning movement in the United States would create a novel opportunity to observe the influence of urbanization on regional climates. Launched in response to the unhealthy conditions of the 19th-century industrial city – environments characterized by widespread disease transmission and heavy industrial pollution – the new towns movement sought to decentralize older industrial cores through the founding of settlements in previously undeveloped rural areas. Although the principal aim of these towns was to relieve population pressures in large cities, the conversion of croplands to urban development would also enable the measurement of climate changes accompanying urban growth over time.

The potential for cities to modify their own climates has been recognized for centuries. During the period of the Roman Empire, for example, ancient city designers noted that streets in Rome became hotter after being widened during the reign of Emperor Nero to accommodate growing commercial activity. To address this concern, it was recommended that streets be "made narrow, with houses high for shade" [4]. Likewise, Middle Eastern cities have long been characterized by narrow, curving, and discontinuous street patterns to impede the penetration of wind-blown sand into residential districts. In the U.S. context, Thomas Jefferson proposed a checkerboard pattern of parks and built-up blocks as the ideal design for a southern city, theorizing that such a pattern would promote air circulation through convection and reduce humidity.

Although early city designers were cognizant of the potential for urbanization to modify temperatures, it was not until the 19th century that appropriate instrumentation was developed to measure the magnitude of such an urban effect. In one of the first books published on the topic of urban climatology, an amateur meteorologist named Luke Howard reported a series of temperature measurements within and around London in 1818. These measurements showed that central London tended to be almost 4°F warmer than the surrounding countryside – the first formal documentation of the urban heat island effect [5].

A wealth of research in the intervening period has demonstrated that four principal characteristics of cities typically render these environments hotter than the surrounding countryside. The first characteristic is a reduction in evaporative cooling brought about through the displacement of vegetation by the streets, parking lots, and buildings of cities. Because the mineral-based materials used to build those facilities are generally impervious to water, a smaller proportion of rainfall is retained by urban surfaces to fuel evaporation. Likewise, the reduced surface area occupied by vegetation limits the quantity of moisture retained for transpiration in plants. As with deforestation in a rural setting, a reduction in moisture availability causes a shift in the surface energy balance from the latent to the sensible heat flux, increasing the quantity of heat released to the air.

A second characteristic of cities conducive to enhanced warming is low surface reflectivity. Due to the darkly hued paving, such as asphalt paving, and roofing materials distributed throughout most cities, a larger quantity of solar energy is often absorbed in cities than in adjacent rural areas with higher surface reflectivity. Unable to compensate for an enhanced absorption of solar energy through an increase in evapotranspiration, a larger percentage of this absorbed energy is returned to the atmosphere as sensible heat and longwave radiation, raising temperatures.

Compounding the problem of diminished reflectivity is the reabsorption of reflected radiation by the vertical surfaces of tall buildings. Because solar radiation reflected by the land surface will travel in all directions, some proportion of this reflected radiation will be absorbed by the surfaces of buildings that block a clear path to the sky. Once absorbed by these surfaces (or re-reflected toward other buildings and the ground), additional heat energy is released into the near-surface atmosphere. The creation of such "urban canyons" in high-density districts thus constitutes a third characteristic of cities that tends to enhance temperatures relative to the surrounding countryside.

Finally, in addition to increasing the quantity of heat energy absorbed and retained from the Sun, urban environments generate copious amounts of waste heat from mechanical processes that can further elevate near-surface temperatures. Such anthropogenic sources of heat – including vehicle exhaust, waste heat emitted from power plants and other industrial operations, and heat mechanically removed from buildings via air-conditioning systems – constitute an important source of heat generation that is far greater within cities than outside of them.

The degree to which these four characteristics of cities elevate temperatures has been the formal subject of climate research since the mid-20th century. Yet, the impact of urbanization on climate can be difficult to measure. For example, in addition to urbanization, changes in elevation directly influence air temperatures. Thus, if an urban weather station is located at a lower elevation than a nearby rural weather station, the urban temperature measurements are likely to be higher than the rural measurements, independent of any differences in land cover. Therefore, the differences in urban and rural temperatures recorded by Howard in the 19th century may not reflect the influence of urbanization alone.

It is for this reason that the emergence of the new towns movement in the early to mid-20th century created an ideal opportunity to directly measure the influence of urbanization on climate. One such new town, Columbia, Maryland, was developed by James Rouse in the 1960s as a social and economic experiment. In designing and building a town from scratch, Rouse believed that innovative patterns of neighborhood design could foster greater social interaction. For example, the clustering of neighborhoods around public schools and small commercial districts, linked together by a network of biking and walking trails, was intended to reduce vehicle travel and to enable residents to live and work in the same small community. Yet, to make sense economically, the new town would need to be sited outside of existing urban areas, where land values were low. Recognizing the value of the proposed town not only for social experimentation but also for measuring the natural growth of an urban heat island, an urban climatologist named Helmut Landsberg began measuring temperatures around the site at the start of the development process.

In an early phase of the town's development in 1968, Landsberg found the maximum temperature differential between locations within Columbia and the surrounding countryside to be about 2°F. During the next six years, the town's population would increase rapidly, growing from about 1,000 to 20,000 residents, and with that growth came substantial changes in the town's climate. By 1974, Landsberg found temperatures in the town's center to be as much as 13°F greater than the surrounding countryside, representing an almost sevenfold increase in the town's heat island [6]. Although such a rate of change during a six-year period is remarkable, more impressive still is the magnitude of warming experienced. The measured rise in temperatures of 13°F is a greater extent of warming than occurred at the global level between the depths of the last ice age and the present period and than is envisioned under the IPCC's worst-case scenario for global warming over the course of the present century. What Landsberg's data demonstrated in the early 1970s was that urbanization was routinely bringing about changes in climate more profound, and over a much shorter period of time, than greenhouse gases had been measured to produce at any spatial scale.

Since the publication of Landsberg's findings, hundreds of studies have measured the existence of an urban heat island effect in all regions of the planet. These studies have employed every imaginable means of measuring urban and rural temperatures, ranging from temperature probes affixed to the roofs of cars to high-resolution thermal sensors aboard orbiting satellites. Indeed, millions of commuters in American cities unwittingly carry out this experiment daily as thermometers integrated into modern vehicles register a declining thermal gradient during a late afternoon drive from city center to suburb. What these studies have uniformly demonstrated is that urbanization elevates temperatures in large cities by a range of 2 to 10°F, with maximum heat islands measured to be in excess of 20°F [7]. Heat islands have been shown to be most intense during periods of clear and calm weather and to increase in proportion with the size of urban populations. Perhaps most important, urban heat islands have been demonstrated to markedly amplify unusually hot days in comparison to nearby rural areas.

Figure 3.4 illustrates the change in heat-island intensity during a prolonged heat-wave event in the Midwestern United States in 1999. During this event, temperatures in excess of 90°F were recorded across Midwestern states for two weeks in July, resulting in several hundred deaths at the height of the heat wave. Similar to data collected during the European heat wave of 2003, urban temperatures were found to be much higher than those in nearby rural areas, with cities consequently suffering the greatest negative effects of the heat wave. However, data from the 1999 event suggest that not only were cities hotter than rural areas, as would be expected as a product of heat-island formation, but also that urban environments were disproportionately hotter. What is revealed in the data from Chicago and St. Louis is that the

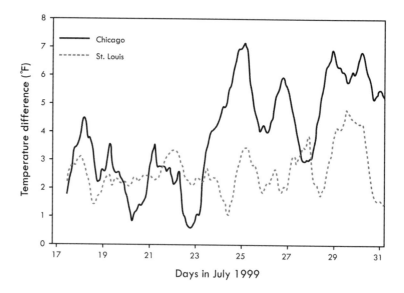

Figure 3.4   Urban–rural temperature difference in Chicago and St. Louis during 1999 heat wave. *Source:* Urban-Rural Temperature Comparison for July 17–31, 1999, image courtesy of the Midwestern Regional Climate Center, Illinois State Water Survey.

difference between rural and urban temperatures almost doubled between the beginning and the height of the heat wave. As shown, a heat-island intensity of about 2–3°F prior to the height of the heat wave was amplified to almost a 5°F increase within St. Louis and to more than a 6°F increase in Chicago. These increments were on top of the enhanced temperatures brought about by the heat wave itself.

The tendency for urbanized areas to amplify heat waves raises a critically important but seldom examined question: To what extent will urbanized areas amplify global warming? Specifically, if the IPCC is projecting a 2 to 12°F increase in globally averaged temperatures by 2100, should we expect a comparable increase at the scale of large cities or an even greater rise in temperatures? The answer to this question reveals a surprising shortcoming in the global science community's approach to climate research.

For anyone who has watched the sitcom *Seinfeld*, the principal locus of planetary climate-change monitoring is a familiar sight. Housed on the floors above "Tom's Restaurant," the eatery frequented by Jerry and friends, is the Goddard Institute for Space Studies (GISS), the division of NASA headed by Dr. James Hansen and responsible for measuring annual

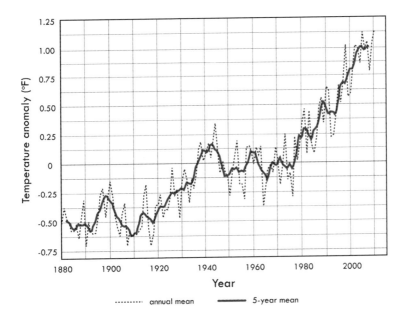

Figure 3.5   Goddard Institute for Space Studies global temperature index.
Source: Goddard Institute for Space Studies, NASA, available at
http://data.giss.nasa.gov/gistemp/.

changes in mean global temperatures. Beyond its little known affiliation
with television lore, the Goddard Institute has received media attention
in recent years in response to its annual release of global temperature
anomaly rankings. Each January, GISS releases its estimate of the mean
annual temperature of the planet as a whole for the preceding year.
It is based on this estimate that the list of the 10 warmest years on
record is compiled each year, with the preceding year inevitably found
to be among the hottest ever measured. Presumably unhappy with this
inexorable march toward higher temperatures, the Bush administration
actively worked to censor the work of Hansen and the GISS during the
early to mid-2000s, serving to further draw media attention to this otherwise obscure outpost of NASA.

The source of the Bush administration's discomfort – the trend in
global mean temperatures updated annually by GISS – is presented in
Figure 3.5. To most effectively illustrate temperature change over time,
temperature trends are reported here as anomalies, which simply represent the difference between the mean global temperature in any year and
a fixed long-term average temperature. What the graph indicates is that
since the late 19th century, when reliable global-scale measurements first

became available, the planet's mean annual temperature has increased by almost 1.5°F. The graph further indicates that 9 of the last 10 years were the hottest years ever measured, a trend inarguably illustrative of the rapid planetary heating underway.

It is important to note that global-scale trends are rarely, if ever, reflective of climate trends experienced at the scale of human settlement. Due to the massive thermal inertia of the oceans (i.e., the ability of the ocean to slowly distribute heat through the depths of its water column), the temperature of the ocean surface should be expected to rise more slowly than the temperature of the land surface, diminishing the rate of change in the global average temperature. Yet, even were we to focus on land-surface temperature measurements alone, the GISS global anomaly data would provide a poor indicator of the rate at which climate is changing in cities. To understand why, we must first consider the methods through which global surface temperature is measured by NASA and other climate research groups around the world.

To measure global-scale temperature trends, climate scientists most commonly make use of networks of meteorological stations positioned across the Earth's surface. One such network, the Global Historical Climatology Network (GHCN), consists of more than six thousand weather stations distributed across every continent and ocean. Given the need for data on weather in proximity to the most heavily populated regions of the planet, a large percentage of the meteorological stations included in the GHCN are located within metropolitan areas. Because urban weather stations are known to be influenced by the urban heat island effect, an overreliance on such stations in estimating mean global temperature may artificially inflate estimates of global climate trends. To address this potential bias in temperature-trend analyses, climate scientists statistically adjust urban observations to mirror nearby rural trends. The effective result of such statistical adjustments is to render the global climate dataset reflective of rural trends only.

This exclusion of unadjusted urban temperature observations from global climate datasets marks a subtle but significant shift from the commonly understood objectives of climate research. By adjusting temperature observations in areas influenced by land-use change, climate scientists are effectively removing the known effects of land-use change from the global temperature record. In so doing, the intended objective of climate research is altered in the following key respect: rather than measuring the degree to which the planet is warming in absolute terms, such analyses measure the degree to which the planet is warming *from greenhouse gases alone*.

What are the implications of an orientation toward greenhouse-induced warming at the global scale? Given the small percentage of the global land surface occupied by cities – estimated at about 3% [8] – the adjustment of temperature observations from urban areas in these analyses likely has little effect on climate trends measured at the global scale. Yet, at the local to regional scale, focusing only on greenhouse-induced warming greatly underestimates temperature trends underway, given that land-use changes in the form of deforestation and urbanization have been shown to impart a strong influence on climate at these scales.

Incorporating modified urban data only, global-scale climate trend analyses carried out by GISS and other global climate research groups provide little insight into the pace and extent of climate change underway in urban environments. At the same time, no supplementary analyses are regularly conducted by these organizations to assess the magnitude of warming in large cities. The outcome of an approach to climate-change monitoring that effectively ignores the most heavily populated regions of the planet betrays an irony seemingly worthy of a *Seinfeld* skit: ask a climate scientist how rapidly the planet is warming and you will get an answer; ask a climate scientist how rapidly your city is warming and you will get a shrug.

Measuring the pace of climate change at the global scale is, of course, a critically important task. Because many drivers of climate operate over large geographic areas, evidence of mean changes at the global scale provides perhaps the most important single measure of global climate change. But the policy-relevant effects of climate change do not occur at the global scale. Heat waves, severe storms, floods, and crop failures are regional phenomena. Manifested as changes in regional climates, climate change must be assessed at subglobal scales to inform the different climate management policies required in different regions of the planet. To date, however, fundamental questions pertaining to the nature of climate change at a range of subglobal scales – particularly below the continental scale – have received only limited attention by the leading climate research institutions such as the IPCC. Chief among these questions concerns the magnitude of climate changes underway in the most heavily populated regions of the planet – large cities.

A key planning challenge confronting the most densely populated regions of the planet today concerns the extent to which rising levels of heat will threaten human health in the coming decades. Extreme heat presently accounts for more weather-related deaths per year than any other form of extreme weather; in fact, extreme heat is more deadly than all other forms of extreme weather combined [9]. Although global-scale climate change is projected to bring more intense storm events and extensive flooding due to rising sea levels, it is quite likely that the rising level of heat at the root of these problems will itself constitute the greatest threat to human health in cities. Thus, there is a critical need for data on the pace of warming trends underway at the urban scale.

To address this need, the Urban Climate Lab at Georgia Tech has developed an urban-scale temperature-trend analysis complementary to NASA's global-scale analysis. Using a subset of the same weather-station data employed in the NASA global temperature analysis, our lab seeks to address a basic question: *Are large cities warming more rapidly than the planet as a whole?*

A key task in answering this question is to distinguish urban climate trends from rural climate trends. If rural weather stations are located far from land-development activities, such as highways, dense buildings, and parking lots, any warming trends detected in those areas are presumably a product of the global greenhouse effect. Cities, by contrast, are subject to the same regional influence of the greenhouse effect at work in nearby rural areas, as well as to the warming influences of dense land development. Separating urban from rural weather stations in measuring climate trends over time thus enables a comparison of warming trends in landscapes subject to different combinations of climatic drivers.

The determination of which landscape types constitute "rural" or "urban" has been a central concern of the field of urban climatology for many decades. In the earliest comparisons of urban and rural climates, when the distance between urban settlements was generally greater than it is today, traveling a fixed distance from a city center into the countryside was often a sufficient basis for distinguishing an urban from a rural landscape type. In recent decades, data on population densities commonly have been used to classify areas by development type. In the context of urban-heat-island measurement, however, areas of low population density may still effect a significant influence on heat-island formation if they have extensive infrastructure development, such as may be found on a remote military base.

The increasing availability of commercial satellite imagery over the last few decades has greatly advanced the precision of urban classification. Night-light intensity, as measured from space, is widely regarded as a reliable means of classifying the extent to which a particular location is urbanized. In what is now probably a familiar image to most readers,

Figure 3.6 shows the intensity of night light associated with urbanized regions worldwide. Through the classification of night-light intensity, highly urbanized regions (bright) can be distinguished easily from peri-urban (dim) and rural (dark) locations [10]. In our work at the Urban Climate Lab, we make use of such night-light intensity data, coupled with information on population density, to identify weather stations in urban and rural areas.

Through this approach, we have measured the difference between urban and rural temperature trends for 50 of the most populous U.S. metropolitan regions during a five-decade period dating back to the 1960s. In each region, average annual temperatures have been compiled for both urban and proximate rural weather stations. The results of this analysis are presented in Figure 3.7. Similar to the GISS analysis of global temperature trends, we measure change in average temperatures over time as anomalies, or deviations from a long-term trend. In contrast to the NASA analysis presented in Figure 3.5, our results are presented as two trend lines, one representing the mean temperature trend of rural weather stations and a second representing the mean temperature trend of urban weather stations.

Three basic conclusions may be drawn from Figure 3.7. First, urban areas, on average, are measurably warmer than rural areas. The vertical distance between the two trend lines is direct evidence of the existence of the urban heat island effect, the mechanism through which land-use change and waste-heat emissions in cities increase temperatures. When measured as an annual average, the difference between urban and rural temperatures across the 50 regions is about 1.5°F. This relatively modest difference in mean annual temperature masks what is a much larger temperature difference on a typical summer afternoon. This is because the intensity of heat islands varies by time of day and season. Thus, averaging over all periods of the day and all months of the year has the effect of greatly reducing the maximum daily differences.

A second important finding highlighted by the two trend lines concerns the relative importance of the two warming mechanisms at work in urban areas. The zero anomaly line in the graph denotes the long-term average temperature for the rural stations included in the analysis, measured during the period of three decades. Any years in which the rural trend falls below this line represent periods of relative cooling, whereas any years in which the rural trend rises above this line represent periods of relative warming. Based on this simple relationship, the rural stations are experiencing a predominant warming trend over the long-term average of about 1.1°F. This number approximates the magnitude

Figure 3.6 Global night-light intensity. *Source:* NASA http://visibleearth.nasa.gov/view.php?id=55167.

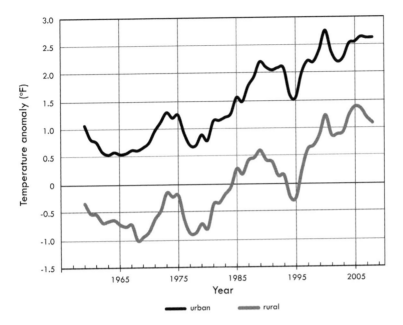

Figure 3.7   Urban and rural temperature anomalies in large U.S. cities: 1961–2010. Data reported as 5-year means for 50 of the most populous U.S. metropolitan regions. *Source:* Updated from B. Stone, Urban and rural temperature trends in proximity to large U.S. cities: 1951–2000, *International Journal of Climatology,* 27, 1801–1807, 2007.

of warming found for the planet as a whole over this period, as illustrated in Figure 3.5, and is attributable to rising concentrations of greenhouse gases over time because areas of significant land-use change have been excluded through the selection of rural weather stations.

The vertical distance between the zero-anomaly line and the urban-trend line represents the magnitude of warming over a rural base-line as a product of two climate-change mechanisms: the global green-house effect and the local heat-island effect. Because the regional green-house gas concentrations that are at work in rural areas are also at work in nearby urban areas, the area under the rural-trend line and above the zero-anomaly line represents the magnitude of greenhouse-induced warming underway in both the rural and urban areas. Thus, the relative contribution of the urban heat island effect is represented most directly by the distance between the urban- and rural-trend lines.[1]

---

[1] With respect to this second conclusion, it should be noted that cities may also experience an enhanced warming effect due to a greater concentration of

What is important to highlight about these relationships is that the magnitude of warming attributable to the urban heat island effect – the vertical distance between the rural- and urban-trend lines – is greater during the five-decade period than the magnitude of warming attributable to the global greenhouse effect – the vertical distance between the zero-anomaly and rural-trend lines. Put more simply, *in cities, land-use change and waste-heat emissions are playing a more significant role in ongoing warming trends than greenhouse gas emissions.*

The final key conclusion revealed by this analysis pertains not to the distance between the urban- and rural-trend lines but to their slopes. By measuring the change in the urban- or rural-trend line from year to year, we can determine how rapidly temperatures are rising, on average, in each environment. During the 50-year period, rural temperatures are shown to be increasing at an average rate of $0.28°F$ per decade. Thus, were this rate of change to continue over the course of the present century, we would expect rural areas to experience an increase in temperature by 2100 of $2.8°F$. Urban areas are found to be increasing at a higher average decadal rate of about $0.42°F$, or a bit more than $4°F$ per century.

When compared to IPCC projections for global-scale warming by 2100, these numbers fall on the low end of the projected range of 2 to $12°F$. Yet, they do not tell the whole story. It is important to note that the IPCC projections account for increasing global concentrations of greenhouse gases over the course of the present century, whereas these urban and rural warming rates are reflective of greenhouse gas concentrations and levels of urbanization consistent with the last 50 years only. Were we to account for the ongoing rapid rise in emissions and continuing urbanization, these projections would be much larger.

A more useful means of interpreting these numbers is to assess the extent to which rural warming trends are being amplified in urban environments. The average decadal rate of warming in large U.S. cities during the last five decades is found to be greater than the rural rate by a factor of 1.5, suggesting an average rate of amplification of 50%.

greenhouse gases in their near-surface airsheds. Although greenhouse gases such as CO_2 are generally considered to be globally diffuse, in that they are widely dispersed around the planet through atmospheric mixing, studies have found higher concentrations over cities, a phenomenon referred to as "CO_2 domes." Yet, studies focused on the extent to which higher concentrations of CO_2 in urban airsheds are contributing to enhanced temperatures have found any such effect to be very limited, measured in Phoenix, for example, to be about $0.2°F$ – a small component of the 5 to $10°F$ enhancement attributed to the urban heat island [11].

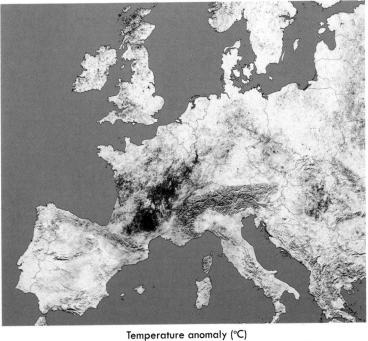

Temperature anomaly (°C)

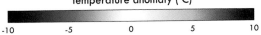

-10   -5   0   5   10

Figure P.1   Temperature anomalies (degrees above or below normal) across Europe on July 31, 2003. *Source:* Adapted from the NASA Earth Observatory http://earthobservatory.nasa.gov/NaturalHazards/view.php?id=11972.

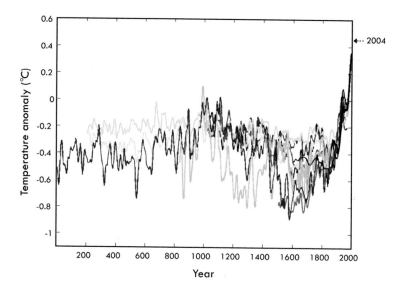

Figure 1.5   Reconstructed global temperature trends. Each curve represents a different temperature proxy; black line represents temperature observations. *Source:* Adapted from R. Rohde, Global Warming Art. For more information on source data, see http://www.globalwarmingart.com/wiki/File:2000_Year_Temperature_Comparison_png.

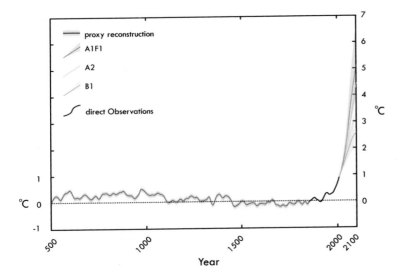

Figure 1.6   IPCC global temperature projections. *Source:* Adapted from I. Allison, N. Bindoff, R. Bindschadler, et al., *Copenhagen diagnosis: Updating the world on the latest climate science*, University of New South Wales Climate Change Center, 2009.

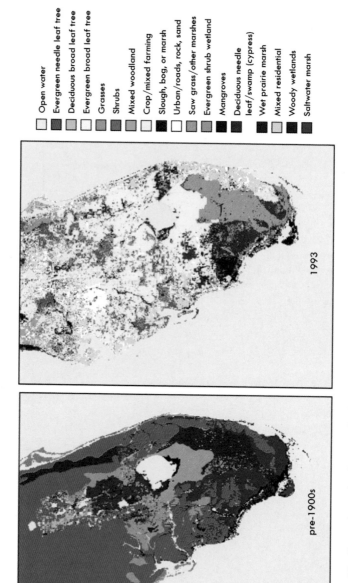

Figure 2.1 Pre- and post-development landscapes in southern Florida. *Source*: C. Marshall, R. Pielke & L. Steyaert, Crop freezes and land use change in Florida, *Nature*, 426, 29–30, 2003.

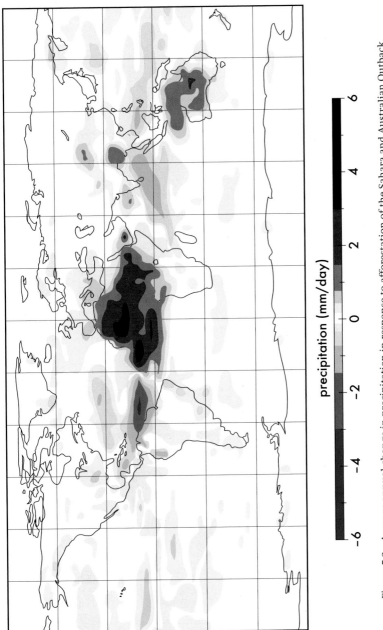

precipitation (mm/day)

Figure 5.2   Average annual change in precipitation in response to afforestation of the Sahara and Australian Outback.
Source: Adapted from L. Ornstein, I. Aleinov, & D. Rind, Irrigated afforestation of the Sahara and Australian Outback to end global warming, *Climatic Change*, 97, 409–37, 2009.

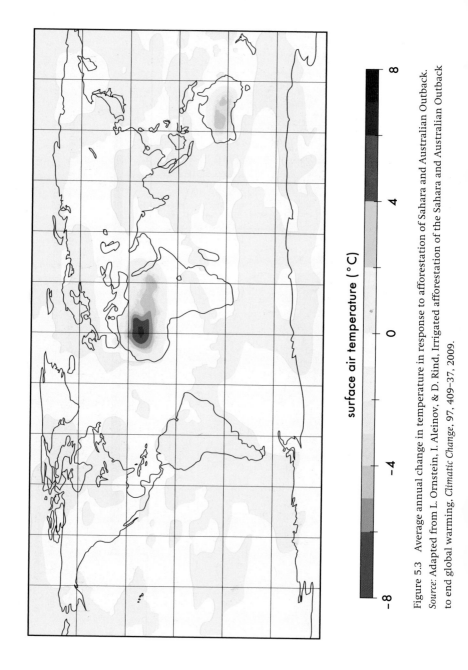

surface air temperature (°C)

Figure 5.3 Average annual change in temperature in response to afforestation of Sahara and Australian Outback.
Source: Adapted from L. Ornstein, I. Aleinov, & D. Rind, Irrigated afforestation of the Sahara and Australian Outback to end global warming, *Climatic Change,* 97, 409–37, 2009.

Therefore, large U.S. cities are experiencing, on average, a rate of warming that is about 50% greater than nearby rural areas. Because the sample of rural weather stations included in our analysis is warming at the same general rate as the planet as a whole, we can conclude that large U.S. cities are also warming at about 1.5 times the rate of global-scale warming during this period.

However, most large cities in our sample are amplifying background, rural rates of warming by considerably more than this average rate of 50%. Figure 3.8 illustrates the average rate of change in urban heat island intensity during the 50-year study period. In cities depicted with a black circle, urban heat islands tend to be expanding for the simple reason that urban temperatures are rising more rapidly than rural temperatures. In some cities, however, heat islands remain unchanged or are shrinking, as indicated by an open circle. In these cities, urban weather stations are still reporting higher temperatures than rural weather stations, but rural temperatures may be rising more rapidly over time than urban temperatures, serving to reduce rather than increase the difference between urban and rural trends. Although a number of factors could be contributing to shrinking heat islands, most commonly this trend is associated with slow-growing cities, many in the Rust Belt, where declining populations may be allowing urban tree canopies to rebound over time.

If we focus only on those cities in which heat islands are growing over time, the pattern underway in more than 70% of the study's cities, we find that urban temperatures are growing at an average rate of 0.56°F per decade, compared to a rural rate of 0.28°F per decade. These numbers suggest that in the majority of large U.S. metropolitan areas, urban temperatures are increasing at a rate 100% greater than proximate rural areas. Thus, most large U.S. cities are not only warming faster than the planet as a whole, they are warming at *double* the rate of global climate change.

Numerous studies focused on the rate of warming at the urban scale have found the pace of climate change in cities to be far outpacing rural warming trends. As to be expected, the most accelerated rates of warming are observed in the most rapidly growing cities. Since the period of the 1960s, for example, Beijing has been warming at a rate more than five times greater than proximate rural areas [12]. Likewise, Tokyo has been warming at 5 times the rural rate for at least 100 years [13]. In Beijing and Tokyo, as much as 80% of the warming underway is a product of urbanization alone, with the remainder attributable to the global greenhouse effect. In large cities of South Korea, urbanization accounts for two-thirds of recent warming trends [14].

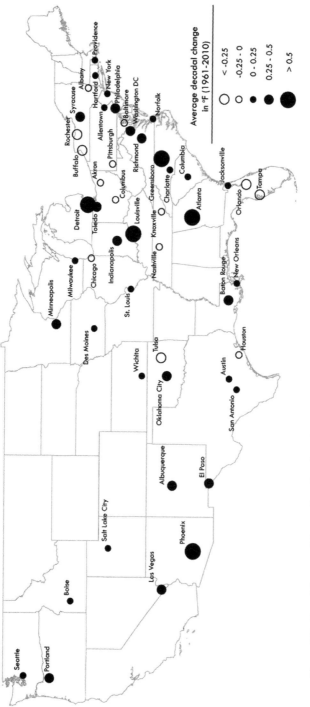

Figure 3.8 Decadal rates of change in urban heat island intensity: 1961–2010. Trends are reported for the 50 most populous U.S. cities for which continuous temperature data are available during this period. *Source:* Temperature-trend data obtained from the Goddard Institute for Space Studies, NASA.

Urban warming in excess of rural- or global-scale warming trends is not limited to large cities of the United States or Asia. Urban temperatures in Madrid, Spain, for example, were found to be increasing at a rate 60% higher than that of nearby rural areas[2] [15]. The rate of growth in London's summer heat island over the latter half of the 20th century was roughly equivalent to the global rate of warming attributable to the greenhouse effect during this period [16]. Many of the largest urban areas worldwide, including the most populous regions of North America, Europe, and Asia, have been found in recent decades to be warming at more than twice the rate of the planet as a whole[3] [17].

The clear conclusion to be derived from these studies is that, in the most populous regions of the planet, land-use change is often playing a greater role in ongoing warming trends than the emission of greenhouse gases. Yet, despite this fact, climate-management policies to date almost uniformly emphasize emissions controls over management strategies focused on the physical characteristics of land use – strategies I refer to as "land-based mitigation." Due to, at least in part, the failure of the IPCC and other climate-science organizations to sufficiently highlight the effects of land-use change on climate, most urban policy makers are preparing for an extent of warming (if they are preparing at all) equivalent to global-scale projections. This assumption profoundly underestimates the magnitude of warming that urban populations are confronting.

No crystal ball was needed to envision what climate change entails for Los Angeles on a late September afternoon in 2010. A few minutes after noon on September 27, a city known more for its temperate, Pacific-cooled climate than for its blistering heat set an all-time temperature record of 113°F. In truth, the magnitude of the heat that day will never be known because the thermometer registering the record succumbed to the extreme temperatures shortly after measuring the new high. The intense heat sparked blackouts for tens of thousands of Los Angelenos, shut down water service to parts of the city, and slowed trains due to concerns over derailments from warping tracks. Yet, perhaps more remarkable than the intensity of the heat that day was its timing, falling just a few days shy of October. Although the previous all-time high in Los

---

[2] Note that the urban and rural temperature trends reported in [15] are for overlapping time periods.

[3] Urban warming trends reported in [16] and [17] are compared to mean global temperature trends during the same periods of observation.

Angeles had also broken the 110°F barrier, this earlier record occurred in late June, the month in which the Sun's annual intensity is at its height. Among the largest cities of the United States, only one now holds a high-temperature record in the month of September. But more such anomalous events are soon to follow.

Data on heat waves suggest that the frequency of extreme heat is increasing rapidly, a trend that is likely to lengthen the season for dangerously hot days as well. As attested to by recent experience, the occurrence of extreme temperatures in cities is not a future event but rather a present manifestation of accelerating climate change. In the coming years, temperatures in excess of 110°F in Los Angeles and many other large cities around the world will become increasingly common, with significant implications for urban infrastructure and human health.

A long-term study of urban development and extreme heat in Shanghai, China's largest city, is representative of trends underway in many large cities around the world [18]. In concert with a more than 50% increase in the region's population between 1975 and 2004, the average intensity of Shanghai's heat island grew from about 0.7°F to 2.8°F, a four-fold increase. As a result of this rapid growth in the city's heat island, the number of extremely hot days, defined as days in which temperatures exceed 95°F, has been increasing as well. On average, the number of extremely hot days experienced in Shanghai is increasing by 0.6 days per year, suggesting an occurrence of about six additional extreme heat days per decade. Thus, if this trend continues during a 50-year period, the number of extremely hot days can be expected to increase from an average of 11 per year today to more than 40. If we assume that the local and global climate-forcing mechanisms at work in China will become even more pronounced during the next several decades, the actual number of extreme heat events experienced per year is likely to be much higher.

What is most instructive about Shanghai's changing climate is that the growing frequency of extreme heat is being produced largely by the city itself. Increasingly, extreme temperatures result less from large-scale weather systems affecting an expansive geographic area and more from local-scale warming mechanisms at work only in the metropolitan area. For example, in 1983, more than 80% of the extreme-heat events experienced in Shanghai were also experienced in the rural hinterland around the city, suggesting the workings of a regional-scale weather pattern. By 2003, less than 30% of the extreme-heat events experienced in Shanghai were also experienced outside the city. Increasingly over time, Shanghai is producing its own weather.

As a result of these trends, the rate of increase in extreme heat events is much greater in Shanghai than in nearby rural areas. Data from rural weather stations in proximity to the city show that the frequency of extremely hot days is increasing by 0.3 days per year, or about three days per decade, compared to six per decade in Shanghai. At this rate, rural areas not subject to Shanghai's heat island are expected to endure only one-half the number of extreme-heat events as the city in 50 years.

More problematic than the total number of extremely hot days is the growing tendency for such days to cluster into heat waves. As observed during the European heat wave of 2003, it is the succession of extremely hot days, rather than the isolated intensity of heat itself, that results in the most heat-related deaths. Also similar to the European experience, data from Shanghai show that the frequency of prolonged heat waves is far greater in urban areas than outside them. During the 30-year study period, rural areas outside of Shanghai experienced an average of 5 heat waves of 5 days or more, whereas Shanghai experienced 18. Five heat wave events lasted for more than ten days in the city, whereas not a single heat wave persisted for that long outside the city.

This disparity in heat-wave duration between urban and rural areas is measured directly in the loss of life. More residents of cities perish in heat waves than residents of rural areas, not because of the larger number of residents in cities but rather because of the disparity in the intensity and length of heat exposure. This fact was made clear during an intense heat wave in 1998, during which the rate of heat-related mortality per 100,000 residents was about four times greater in the city than outside it – a ratio mirroring heat-wave outcomes in many regions of the world. What the recent history of Shanghai and other large cities of the world clearly demonstrates is that the physical structure of the urban environment itself is often a more powerful driver of climate change than the global greenhouse effect. In combination, these two climate-forcing mechanisms will require in the coming decades fundamental changes in the design and management of cities.

So, how high will urban temperatures climb? Although the answer to this question is likely to vary considerably by geographic region, our analysis of historical trends in large U.S. cities provides a reasonable baseline for assessing the magnitude of future warming in general terms. Based on these trends, urban heat islands were found to be amplifying global-scale warming trends by an average of 50% across large U.S. cities. If we assume that this amplification rate will remain constant during the coming decades, we can get a sense of the likely magnitude of warming, absent significant local and global mitigation efforts.

Taking my own city of Atlanta as an example, summer temperatures are likely to increase substantially over the course of the present century. If we consider global-scale warming trends alone, the IPCC is projecting an increase in average temperatures of between 2 and 12°F planet-wide as a product of greenhouse-induced warming by the year 2100. As the growth in global greenhouse emissions has been tracking a worst-case scenario since the 1980s, a conservative point estimate for global-scale warming during this period is 7°F, the midpoint of the IPCC range. Using this estimate, if Atlanta is subject to the same general degree of warming found globally over the course of this century, we might anticipate the city's high temperature of 89°F for a typical July afternoon to increase to 96°F. This number itself marks a dramatic shift in climate because Atlanta has exceeded 100°F on only five occasions since temperatures were first recorded in the 19th century – with all of these days occurring since 1980.

Of course, Atlanta, like other rapidly growing metropolitan regions, should expect the urban heat island effect to further amplify global-scale warming trends. If the degree of heat-island amplification in Atlanta is consistent with that of all large U.S. cities (i.e., 50%), a statistic that reflects trends in cities with both growing and shrinking heat islands since the 1960s, Atlanta would expect to see temperatures climb by an average of about 11°F, bringing a typical July afternoon up to about 100°F. Yet, because Atlanta's heat island has been intensifying over time, the amplification of background, global-scale warming trends during recent decades has been considerably higher than the national average. Applying the higher rate of amplification found for warming cities (100%) to the IPCC projection of 7°F, the expected temperature for a typical July afternoon climbs to an alarming 103°F.

As a back-of-the-envelope calculation, this estimate of future climate in Atlanta may fail to account for important climate-related feedbacks that could dampen or further amplify future temperatures. In some instances, for example, intense heat islands have been found to induce rainfall, which could serve to moderate regional temperatures if sufficient water vapor is available to fuel frequent showers. It is just as likely, however, that such feedback mechanisms would work in the other direction, serving to amplify temperatures. For example, as high temperatures routinely surpass 100°F in Atlanta during the next few decades, the extent of the region's dense urban forest will be diminished. Largely composed of mid-latitude, temperate forest species, such as oak, maple, and poplar, much of Atlanta's present tree canopy is likely to be lost in response to prolonged exposure to temperatures in excess of their cooler and more stable historic environments. Indeed, evidence suggests this process is already underway, with tree canopy in Atlanta neighborhoods found to be declining during the most recent decade. As this forest cover is lost, Atlanta's primary defense against rising temperatures is being lost as well.

The extremity of urban warming projected for Atlanta is well supported by modeling studies focused on future warming trends in large cities. The most comprehensive of these studies to date quantifies the influence of a doubling of atmospheric carbon dioxide coupled with anticipated population growth by 2050 for all urbanized regions of the planet [19]. By accounting for the effects of both urban land-use change and growing emissions of waste-heat energy from population growth, climate scientists are able to model the combined effects of greenhouse gases and urban heat island effects on cities. In good agreement with historical warming rates found in our study of U.S. cities, this study finds cities to be amplifying greenhouse-induced warming effects by between 50% and 100% in most urbanized regions of the planet, resulting in a far greater magnitude of warming in cities than outside of them.

Such a dramatic increase in temperatures will be manifested in a rapidly growing number of extreme-heat events, particularly during the nighttime hours when residents are most susceptible to the health effects of heat. The estimated number of extreme-heat events for some of the world's most populous cities is troubling. Los Angeles, for example, would see the number of such events increase fourfold from about 10 today to more than 40 in 2050. Sao Paulo, Brazil, a city in which millions of residents lack air conditioning, would experience a more than 1,500% increase in extreme-heat events, rising from less than 5 today to more than 80 by 2050. More troubling still are the estimates for Lagos, Nigeria, which may see the number of extreme-heat events increase from less than 10 today to more than 150 during the next four decades. This finding merits reflection: on average, every third night in Lagos may exceed the threshold by which heat-wave temperatures are defined today.

Were such climate extremes likely to become commonplace during the course of the next few decades, we would expect to find clear evidence of these trends in our current climate – and we certainly do. As demonstrated by the first decade of the 21st century, the inexorable march toward a hotter planet continues in what is approaching a linear trajectory. The 2000s were the hottest decade on record, with the 1990s and 1980s falling into the second and third positions, respectively. The year 2010 was both the hottest year of the 2000s and the hottest year ever measured. At the forefront of these extremes are urban temperatures, which are rising far more rapidly than global temperatures. A cursory examination of the summer of 2010, the most recent summer at the time of writing, provides a wealth of evidence as to just how extreme urban climates have already become.

Although a number of cities, led by Los Angeles, would break all-time temperature records in 2010, the more revealing trend was to be found in the frequency of days in which excessively hot temperatures were reached. In my own city of Atlanta, it is the persistence of hot days, rather than the extremity of any single day, that provides the clearest evidence of the pace at which climate change is progressing. Because Atlanta is situated at an elevation of 1,000 feet and is well forested, its historical climate is more moderate than one might expect given its southern latitude. As a result, maximum temperatures during the month of June are typically in the low to mid-80s, with a historical average of just eight days climbing into the 90s. In June 2010, however, the number of days in the 90s exceeded this long-term average by 140%. Nineteen consecutive days that month exceeded this threshold, each more than six degrees higher than the usual high temperature for the month.

During the course of June, July, and August 2010 in Atlanta, a total of 64 days registered in the 90s, 33 more days than would be expected in a typical year – an entire additional month of hot days. What is perhaps most remarkable about these numbers is that the reference period for determining "typical" temperatures for these months is 1971 to 2000, a period during which urban temperatures were already rising steadily. The number of unusually hot days in Atlanta during the summer of 2010 was more than double that of an already hot reference period. And Atlanta was not alone.

In June 2010, Washington, DC, experienced 2.5 times as many days in the 90s as in past years, with the number of such days increasing by about 70% during the course of the entire summer. The dramatic doubling of expected days in the 90s observed in Atlanta was repeated in a large number of cities in the summer of 2010, including Charlotte, Louisville, Miami, and Philadelphia, among others. New Yorkers suffered through the city's hottest summer in history, with more than three times the normal number of days in the 90s. In Dallas, residents rarely ventured outdoors during a spate of 18 consecutive days with temperatures topping 100°F.

In climatological terms, these numbers are astounding. Were 2010 to be a statistical aberration, a highly anomalous year that would remain an isolated high point on the global-temperature-trend graph, we might be hesitant to interpret it as indicative of a profound environmental shift. Yet, 2010 will prove to be no more anomalous than were 1988, 1990, 1995, 1997, 1998, or 2005, previous record-setters whose reign was all too short-lived. The pace at which cities are heating up today makes it entirely probable that the untenable urban conditions foreseen by climate models in the near future will come to pass. If 2010 proves to be anomalous, it will be only in the most ironic sense: the hottest year yet measured will be among the coolest you will live to see again.

The August 11, 2010, match-up between the Texas Rangers and New York Yankees was a sweltering affair. Hosted by the Rangers outside of Dallas, the baseball game took place in the middle of Dallas's string of excessively hot days that summer, with afternoon temperatures climbing to 105°F. To make the experience of an open-air arena in Texas a bit more tolerable, the team owners many years ago resorted to scheduling all of the Rangers' home games in the evenings during the summer. However, the late scheduling provided little relief that night, with temperatures still hovering around 100°F at the start of the game. The Rangers' pitcher was said to look exhausted on the mound, an observation supported by his poor performance, and the Yankees won, 7–6.

Although a hot night in Texas is hardly remarkable, the response of the team's management to concerns over the performance of its starting pitcher was. Soon after the game, the owners acknowledged that they were exploring the option of having thermal shields installed around the stadium to better protect players and fans from the heat. Although the owners' interest in cooling their players may have as much to do with re-signing a star pitcher as anything else, their unconventional proposal speaks to a broader trend: cultural rituals like summertime baseball are already threatened by rising temperatures.

To date, no major league baseball teams have cancelled games due to extreme heat, but a growing number of Little League teams are having to do so. During the summer of 2010, media stories on the rash of heat waves recounted an increasing reluctance on the part of city dwellers to venture outdoors, particularly for activities involving children. As typical summer-afternoon temperatures continue to creep from the 90s into the 100s in large U.S. cities, our daily lives can be expected to change in innumerable, even if unacknowledged, ways. Summertime sports will be increasingly played indoors, outdoor festivals and concerts will become less common, urban gardening will grow far more challenging, and enjoying an evening meal at a sidewalk café will become a less appealing option, to name just a few examples.

Of course, the loss of such cultural rituals marks only the most trivial of consequences associated with climate change. Yet, it may be in these subtle changes that the magnitude of the shift underway is made most apparent. Climate change is not simply manifesting itself through the periodic extreme event; it is starting to alter the rhythms of everyday urban life. Recent experience finds these changes arriving in our lives not with a bang but with the whimper of increasingly uncomfortable conditions. And it is during this period between the uncomfortable and the soon-to-be intolerable that cities must fundamentally remake themselves to remain viable enterprises.

It has been my purpose in this chapter to examine the ways in which cities are directly influencing their own climate and to consider the implications of amplified warming in these most populous regions of the planet. In the book's final two chapters, I explore the policy options available to governing institutions charged with safeguarding urban populations from the most detrimental effects of climate change. Their task is immense. Yet, just as the physical characteristics of cities are amplifying the rate at which urban environments are warming, these characteristics can be modified to strongly counteract these trends. It is to this topic that we now turn our attention.

4

The Green Factor

It is sometimes observed that all explorations of climate change lead back to the Sun. Whether produced from a shift in the surface energy balance or the release of ancient carbon through fuel combustion, the excess thermal energy that now confronts us comes ultimately from the Sun. Of the innumerable pathways through which the Sun's energy warms the environment, human respiration is undoubtedly low on the list of climate-change agents. Yet, in cities at least, the extent to which heat produced through the human metabolism of food is influencing ambient temperatures is now measurable, accounting for almost 5% of the total waste-heat burden in some cities.

Although the dense concentration of human bodies has always influenced micro-climates to some extent, that such a seemingly finite biological process could generate a warming signal at the scale of urbanized regions is wholly a reflection of just how populous these regions have become. Today, urbanized regions are not only home to the majority of the global population, the number of new arrivals to cities each year from births and rural migration is outpacing the number of new arrivals to the planet. As a result of this trend, the next doubling of the urban population will occur in less than 1% of the time required to reach its current size: about 50 years. At that time, around the middle of this century, the total urban population will be equivalent in size to the total global population circa 2004 [1].

A doubling of the urban population during the next five decades will bring with it profound changes to the extent and character of the built environment. The bulk of this population growth is forecast to occur in developing nations, but cities of the developed world will be no less affected by physical change. In the United States, for example, it is estimated that an area equivalent in size to the current national urban footprint will be added or redeveloped in cities during the next

25 years alone, with much of this development already planned [2]. If this growth assumes the pattern of further outward expansion, with new roads, buildings, and further deforestation, urban governments will have managed to compound greatly the climate-related challenges they already confront.

There is reason to expect, however, that the decentralization of cities in response to population growth may soon reverse course. With the release of its annual *World Energy Outlook* report in November 2010, the International Energy Agency (IEA), the organization charged with monitoring the global energy supply, acknowledged for the first time that conventional supplies of oil are in decline. Although the physical limits to the production of nonrenewable sources of energy have been understood for more than a half-century, government entities charged with forecasting energy production, such as the IEA and the U.S. Energy Information Administration, have steadfastly obscured the possibility of exhaustion of the global supply of fossil energy in their annual forecasts for several decades. Indeed, the 2010 report marked not only one of the first instances in which the phrase "peak oil" appeared in an IEA forecast, its appearance also coincided with an acknowledgment that the event had already occurred four years earlier, in 2006. With this single report, the concept of peak oil graduated at once in the government literature from a fringe theory unworthy of mention to an historical event for which it is now too late to prepare.

Although the peaking of the conventional oil supply is likely to alter the structure of urbanized regions in profound ways, its implications for combating climate change remain uncertain. On the one hand, a tightening of global energy supplies may coincide with a sooner-than-expected peaking of global greenhouse emissions, assuming a declining supply is met with enhanced conservation and an accelerated transition to renewable sources of energy. On the other hand, the likely substitution of more carbon-intensive energy sources for transport fuels, such as coal and tar sands, could greatly augment emissions in the coming decades. In either instance, the means through which climate change is to be managed in an energy-constrained world must themselves be less energy intensive. It is with respect to this criterion that land-based mitigation strategies hold great promise.

It is my intent in this chapter to explore the options available to urban governments to offset the magnitude of warming underway as a product of both the global greenhouse effect and, most important, land-surface changes that are enhancing heat absorption and retention in cities. None of these options will reverse long-term warming trends

but, if pursued aggressively, combinations of these strategies can slow
the pace of warming over time and substantially moderate the extremity
of climate events – events not limited to extreme heat alone. Although
cities in different climatic regions require different approaches, all management approaches are born of a common theme: land-surface changes
are the single most effective option available to cities to counteract the
very real threats of climate change during the next half-century.

Against the backdrop of surging urban populations and declining global
energy supplies, the potential to slow the pace of warming in cities
through a reintroduction of dense vegetative cover grows ever more
appealing. For among the full suite of conceivable approaches to cooling
urban environments, none is more effective or less energy intensive than
planting trees. Long valued as a means of shading and beautifying cities,
urban trees assumed an almost iconic status during the heyday of the
City Beautiful movement in the United States in the late 19th and early
20th centuries, a period during which most major cities followed New
York's lead in developing extensive urban parks and networks of street
trees. Employed more as an architectural juxtaposition to the sterility
of the industrial city than as infrastructure, trees and other vegetation
nonetheless endowed urban areas for several generations with an array
of ecological services. Yet, the cycle of disinvestment in cities occurring
during the latter half of the 20th century would contribute to a hollowing
out of both the built and vegetative fabric of urban cores.

My own city of Atlanta provides a characteristic example of the
boom-and-bust cycle of tree cover in many American cities. During its
own foray into the City Beautiful movement of the late 19th and early
20th centuries, Atlanta would see several of its neighborhoods designed
directly by the firm of Frederick Law Olmsted, one of the two designers
of New York's Central Park, or based on Olmstedian principles. As con-
veyed by their names, the neighborhoods of Druid Hills, Ansley Park,
and Inman Park make use of rolling topography, ample greenspace, and
densely planted trees along gently curving streets to create a sense of
the pastoral within the boundaries of a rapidly growing urbanized area.
Yet, much like their Central Park prototype, these neighborhoods were
not originally carved from heavily forested tracts of land but rather
were wholly engineered to achieve this effect through the planting
of thousands of trees in parks, along roadways, and across residential
lots. Shown as mere saplings in the earliest photographs from these
neighborhoods, the various hardwood species of oak, maple, and poplar
would grow over time to establish an unbroken canopy over many areas,

consummating the intended aesthetic while greatly moderating the environmental stresses of the city.

Had these trees been planted in a rural setting, this first generation of cultivated forest feasibly could have remained intact for well more than a century. Yet, as the inhospitable conditions of an urban setting, including air and water pollution, introduced pests, and the vagaries of neighborhood "improvements," have taken their toll, these tree canopies have been substantially thinned in recent decades. Perhaps most important, a periodic schedule of replanting needed to maintain these canopies has not been followed during periods of economic decline. As a result, satellite images show reductions in these legacy canopies of more than 10% in some neighborhoods in last decade alone. With the loss of this tree cover comes a growing need for engineered infrastructure to replace ecological services long provided at virtually no cost.

It is with respect to this need for engineered infrastructure that the value of urban tree cover is revealed most clearly. Between the mid-1970s to the mid-1990s, the Atlanta region lost almost half of its tree canopy, mostly to sprawling suburban development [3]. Such immense deforestation in and of itself is no small feat and was a direct manifestation of a continuing shift in regional populations from urban center to periphery playing out across the country at that time. In contrast to earlier periods of rapid land-cover change, these trees were not lost through a succession of forest to cropland but rather were displaced by the most hydrologically inert materials imaginable: concrete, asphalt, and roofing shingle.

As a consequence of this rapid and extensive deforestation, a raft of ecological services have been lost in and around Atlanta, including the ability of regional tree cover to regulate stormwater runoff. Because trees are highly effective at aerating soils through plant respiration, the ability of once-forested landscapes to infiltrate rainwater through porous soils is greatly diminished, particularly when deforested zones are resurfaced with buildings, roads, and parking lots. Unable to infiltrate to groundwater systems, a much greater percentage of rainfall must now be collected and diverted to engineered stormwater systems. In some cities, the stormwater and sanitary sewer systems underlying streets run through separate pipes, with stormwater diverted directly to receiving streams, rivers, or lakes and sewage transported to wastewater treatment plants. However, in most cities, the less costly alternative of designing a single network of pipes to manage both sewage and stormwater was pursued in the late 19th and early 20th centuries, establishing an infrastructure pattern unsuitable for the physical growth that would follow.

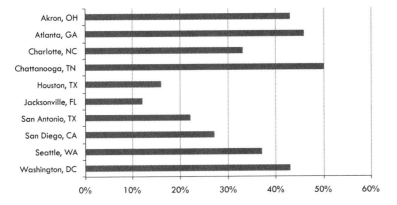

Figure 4.1   Percent decline in tree canopy of large U.S. cities. All trends
are reported for a period of at least 10 years between 1973 and 2008.
Source: American Forests.

Before passage of the federal Clean Water Act in 1972, a ready
solution to the principal design flaw of such combined sewer systems was
easily found: in times of high system volumes during heavy-rain events,
sewage-laden stormwater simply would be permitted to spill into natural
streams and rivers. Yet, so great would Atlanta's stormwater burden grow
as the region's forested land cover was whittled away that such overflow
events became entirely routine, with thousands occurring each year in
the 1990s. At present, in response to a federal court order, Atlanta has
at last undertaken the massive task of separating its stormwater and
sanitary sewage systems, a project that will require the better part of
a decade. The cost of this project, designed ultimately to replace an
ecological service once provided at no cost, is estimated to be $4 billion –
enough to reforest again an area many times greater than that of the city
itself.

Atlanta's experience is mirrored to some degree by that of most
large metropolitan regions of the United States. As presented in Figure 4.1, in virtually every large city in which the regional forest canopy
has been measured, the extent of tree cover has fallen dramatically. This
fact, perhaps more than any other, leaves cities highly vulnerable to the
impacts of climate change, for much like the natural bulwark provided
against storm surges by coastal marshlands, forests in and around cities
absorb prodigious amounts of rainwater during storm events. This natural sponge not only limits the volume of stormwater that needs to be
transported by natural and engineered channels but also substantially
slows the rate at which the levels of receiving waters rise, thereby greatly

reducing the potential for flash flooding. As climate change brings storm events of greater frequency and intensity, the magnitude of damage resulting from regional flooding will be inversely proportional to the spatial extent of trees and other vegetative cover.

Arguably more valuable than the absorption of water during rain events is the retention of moisture by trees and other vegetation between periods of rain. As we have seen, moisture availability within cities is often the single most important variable governing the pace and extent of warming over time. In sharp contrast to the issue of stormwater management, no human infrastructure exists to replace this ecological service once it has been lost. For urban governments serious about the public health threats of climate change, reforestation of metropolitan regions, including dense inner-city cores, offers perhaps the least costly and most effective strategy available to manage extreme heat.

In the fall of 2006, when Mayor Villaraigosa of Los Angeles announced that his city would be planting one million trees, it may not have been his intent to throw a gauntlet at the feet of other major U.S. cities, but a trend was at least started. Not about to be outdone by a West Coast rival, New York's Mayor Bloomberg announced less than a year later that his city would plant a million trees as well, augmenting the city's total by 20%. Next would come Houston, relying on regional nongovernmental organizations to take the lead in planting yet another million trees. The friendly competition makes for good politics in an era in which cities are finally outcompeting their suburbs for a modest share of new residents. Yet, more than an investment in neighborhood aesthetics, such initiatives will come to be viewed as a down payment on the massive program of climate management soon to be undertaken by cities – the beginnings of an inland seawall to guard against a rising tide of heat.

That what may be the first attempt by a U.S. city to directly manage climate change on a large scale was undertaken in Los Angeles is no coincidence. Long recognized as the nation's capital for smog, Los Angeles has been the proving grounds for a wide range of ambitious air-pollution control efforts, and heat-island management offers a novel approach to influencing the chemical process through which ozone, the main ingredient in smog, is formed. Rising temperatures intensify ozone formation by increasing the quantity of ozone precursors, including nitrogen oxides and volatile organic compounds, emitted from cars, gas stations, power plants providing electricity for air conditioning, and a range of natural sources. So, whereas reductions in the number of miles we drive or the amount of electricity consumed by air-conditioning systems provide one approach to controlling ozone, moderating ambient temperatures in cities provides another.

Of course, many do not consider changing the weather in cities to be a viable policy option for managing environmental problems. Yet, it should be. The geographically expansive landscape of Los Angeles, spanning an area greater than several northeastern states, has modified Southern California's climate in profound ways, most directly by increasing regional temperatures. Having already increased temperatures by about 5°F on a typical summer afternoon, it is fully within our power to cool Los Angeles down by several degrees through some creative reverse engineering.

With this goal in mind, a group of scientists at the Lawrence Berkeley National Laboratory, at the University of California at Berkeley, has been exploring for more than 20 years the potential for heat-island mitigation to improve urban air quality. Their innovative research provides a powerful roadmap for addressing the now more pressing problem of climate change in cities. Using a set of regional climate and air-chemistry models, the Lab's Heat Island Group has experimented with an array of strategies, including tree planting and the use of reflective roofing and paving materials, to quantify the extent to which proactive measures could abate the region's heat island and improve air quality.

In one such experiment, the Berkeley Lab researchers evaluated the benefits of a massive tree-planting scheme in Los Angeles. In contrast to the million trees currently being planted within the city proper, the researchers modeled the effects of planting 11 million trees throughout the much larger Los Angeles metropolitan region. Through this simulation, shade trees were planted in proximity to houses and buildings, along roads, and in parks across the metropolitan area. The results of this analysis showed that such tree planting alone could reduce summer afternoon temperatures by almost 3°F, diminishing the region's average summer heat island by more than half [4].

To fully appreciate the magnitude of this finding, it is useful to compare the modeled effects of heat-island mitigation in Los Angeles to alternative climate-management strategies. At present, the most commonly advocated approach to managing rising temperatures at both the global and regional scales is a reduction in the emissions of greenhouse gases, as embodied by the Kyoto Protocol and a number of national and regional carbon-management programs. Numerous cities have now developed "climate action plans," designed to identify strategies for reducing greenhouse gas emissions at the urban scale. Los Angeles, for example, has adopted a goal of reducing city emissions of greenhouse gases to levels 35% below 1990 emissions by 2030. This is a laudable goal, to be sure, and represents a climate-change mitigation program far in advance of anything underway at the federal level in the United States. But it must also be acknowledged that even if reached, the attainment of this goal alone would not likely reduce temperatures in Los Angeles by a fraction of a degree – in fact, temperatures would likely increase dramatically by 2030 under this program.

No emissions-control program of any ambition and at any scale, including a globally coordinated effort to eliminate greenhouse gas emissions altogether by 2030, or even by tomorrow, would yield a comparable reduction in temperatures during a typical Los Angeles summer afternoon than the extensive planting of trees. This is precisely the consequence of having delayed significant action on emissions controls for so long: the accumulated thermal inertia associated with historic carbon emissions is sufficient to propel ongoing warming trends, entirely independent of future emissions, for what is now estimated to be hundreds of years [5, 6]. Therefore, although only reductions in carbon emissions coupled with an increase in carbon sequestration can ultimately reverse global climate change over the very long term, at the scale of cities, only land-based mitigation can yield measurable improvements during the period of our own or our children's lifetime.

The good news here is that cities are empowered to effect some measure of control over their own climate fate, at least with respect to extreme temperatures. Indeed, urban reforestation programs have been found to be effective not only in reducing the intensity of day-to-day heat islands but also in significantly moderating extreme-heat events. Returning to my own city of Atlanta, regional-scale climate models have shown that a doubling of the region's forest cover, an ambitious but achievable objective given the region's low density of development, could reduce temperatures on the hottest days by more than 12°F [7]. This is an encouraging finding in that it suggests the potential to substantially counteract even the most intense heat-wave events that are expected to affect the southeastern United States with increasing frequency.

Neither Atlanta nor Los Angeles is particularly unique in the modeled potential for tree planting to counteract warming trends. Most large cities situated in naturally forested regions, a condition met by cities in which the majority of the U.S. population resides, would benefit substantially from a reintroduction of forest cover within and in proximity to the urbanized zone. Studies focused on the climate benefits of increased tree canopy in a large number of U.S. cities, including New York, Philadelphia, Los Angeles, Chicago, Washington, DC, Houston, and Miami, have shown the potential for heat-island reduction to be substantial, depending on the extent to which the canopy can be expanded.

The costs associated with an extensive urban-reforestation program would not be trivial. Based on a tree-planting program sponsored by a California power utility, the Lawrence Berkeley Lab scientists estimated the per-tree cost of planting in Los Angeles to be about $70 in today's dollars. More recent studies have cited costs of $100 or greater [8]. Assuming a planting cost of $100 per tree, the total cost of a program designed to plant 11 million trees throughout the Los Angeles basin would be about $1.1 billion – no small change, to be sure. Yet, balanced against these costs are the direct savings associated with reducing temperatures: avoided electricity costs for air conditioning and avoided costs for pollution controls required to manage heat-island–induced smog formation. These costs are estimated to be more than $600 million a year in today's dollars in Los Angeles, so that even the narrowly measured benefits of extensive tree planting suggest that such a program could pay for itself in a short period of time [4].

The actual benefits of trees, of course, should be measured more broadly than avoided energy and pollution-abatement costs alone. Research on tree planting in cities has documented a wide array of additional benefits, including city beautification, stormwater management, and CO_2 sequestration. Among the most significant and perhaps least appreciated benefits is enhanced property values. Studies comparing similar properties with and without mature canopy trees find that each large front-yard tree can increase property values by an average of almost 1% – representing a very large economic benefit when totaled across an urbanized area. When accounting for the full range of costs, including planting and annual maintenance, versus benefits, including reduced energy and stormwater management costs, enhanced property values, improved air quality, and carbon sequestration, the annual benefits of urban trees have been found to exceed costs by a factor of 1.4 to 3.1 (Figure 4.2).

It is likely, however, that the most valuable economic benefits of urban trees moving forward have been overlooked altogether by these studies. As suggested by both regional climate modeling and observations from heavily and sparsely canopied neighborhoods during Europe's 2003 heat wave, the presence of an extensive urban forest has the potential to substantially moderate the extremes of heat-wave conditions. During these conditions, already increasingly common in large cities, a reduction in temperatures of several degrees can span the threshold between operational and nonoperational infrastructures, including

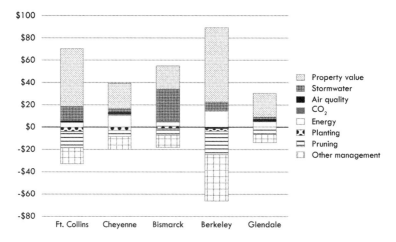

Figure 4.2   Benefits and costs of urban trees in five U.S. cities. *Source:*
Adapted from G. McPherson, J. Simpson, P. Paula, et al., Municipal forest
benefits and costs in five U.S. cities, *Journal of Forestry*, 103, 411–416, 2005.

electrical power plants and rail service, as well as the threshold beyond
which human heat stress becomes epidemic. In a word, urban trees
greatly enhance climate resilience – an attribute that will be regarded as
increasingly invaluable in a warming world.

Manhattan's Lower East Side is not a neighborhood known for its tree
canopy. With its wide streets, generous sidewalks, and endless rows of
five-story walk-ups, the long-standing immigrant enclave yields little real
estate for the planting of trees outside of its few parks. And in this respect,
the neighborhood represents well the island borough as a whole. Studies
have found that only a paltry 3% to 8% of Manhattan's surface is overlaid
by tree canopy, with more than 80% of the island occupied by impervious
cover [9]. It is for this reason that the city's million-trees campaign will
add little new greenery to the Lower East Side: the planting of trees in
such densely developed urban cores requires not only a shovel but also a
jackhammer.

Home to 1.6 million residents, Manhattan will need to expand its
hydrologically active surface if it hopes to moderate the temperature
extremes amplified by its expansive asphalt. A key resource in doing so
will be its roof area. Estimated to occupy about 40% of the borough's
total surface area, Manhattan's rooftops alone provide more potential
planting area than found in the entirety of some modest-sized American
towns. Because the vast majority of these rooftops are flat or low sloping,

there is a great potential to retrofit these elevated surfaces with green roofs, and the evidence available on the climatic effects of such an undertaking is encouraging. A study in Toronto found that the greening of just 5% of that city's area in the form of roof gardens lowered temperatures by almost 1°F [10] – a surprisingly large benefit for the modification of a relatively small surface area.

A roof-greening project underway in the Lower East Side speaks to the potential for an emerging ecological urbanism to galvanize community interest and bring about rapid change. Concerned about the limited opportunities for his daughter to engage nature in the city, Michael Arad, designer of the World Trade Center Memorial, proposed that a working farm be installed atop her elementary school complex – a project that has come to be known as the Fifth Street Farm. First envisioned as a low-tech initiative involving plastic wading pools filled with planting soil, the idea for children to grow their own food atop their school has attracted so much support from parents that it has now graduated to something much larger.

At completion, the roof garden will occupy 3,000 square feet atop the Robert Simon School Complex, home to two elementary schools and a middle school, providing sufficient space for the growing of fruits and vegetables to be consumed in the school's cafeterias. The ability to raise more than $500,000 for the project from both public and private sources attests to a diverse set of interests that urban agriculture can serve: ecological education, a means to improve the quality of food in schools, and an increase in recreational space for schoolchildren [11]. Likely lower on the list of benefits perceived by donors is the potential for a rooftop farm to combat climate change. Yet, as a good idea spreads, it will do just that.

In the less than two years that have passed since the Fifth Street Farm's inception, a growing number of schools in Manhattan have started raising funds and hiring architects for their own rooftop gardens. Although the greening of school roofs alone will not insulate the city from rising temperatures, the development of building-integrated vegetation throughout the city could measurably reduce temperatures in neighborhoods where there is insufficient space for extensive tree canopy. Studies from similarly dense urban areas have found the conversion of 50% of the available roofing area to green roofs to be associated with a reduction in temperatures of as much as 1.4 to 3.5°F, depending on the extent of irrigation [12]. In cities populated by flat-roofed buildings, the resurfacing of a large extent of roof area with planted material is not out of reach. At present, it is estimated that 12% of all flat roofs in Germany are green roofs.

Figure 4.3   Green roof on Chicago City Hall. *Source:* Tonythetiger http://
en.wikipedia.org/wiki/File:20080708_Chicago_City_Hall_Green_Roof.JPG.

The extent to which such roofs can mitigate heat-island formation
is directly dependent on the ratio of roof area to total built area and, more
generally, the stem length of the species of plants incorporated into such
roofs. Known technically as *intensive* green roofs, roofs engineered for a
depth of planting media of 6 inches or more are capable of supporting not
only species of grass and shrubs but also small trees. More common are
roof structures designed to support only a few inches of planting media
distributed over the full extent of a flat roof, referred to as *extensive* green
roofs (Figure 4.3). Extensive green roofs generally make use of highly
heat- and drought-tolerant species of sedum plants, negating the need
for active irrigation. Because such roofs tend to provide a more shallow
reservoir for moisture and evapotranspiration, extensive roofs are likely
to be somewhat less effective than intensive roofs in cooling the ambient
air, but both approaches are promising strategies in the most densely
built cities.

Like street trees, green roofs carry the potential to pay for themselves over time in savings from both cooling and heating buildings
and from increased longevity of roofing materials. In Germany, where
a mature green-roof industry has been established, the cost of installation ranges generally from $8 to $15 per square foot [13]. A higher but

decreasing cost range of $10 to $25 per square foot governs the U.S. industry, where green roofs remain little more than a novelty [14]. Because green roofs not only cool the ambient air over a building but also provide a biologically active stratum of insulation, they are highly effective in moderating heat transfer across the roofing membrane. The use of living plant materials as a barrier between indoor and outdoor environments enables the building to physically respond to changing daily and seasonal weather conditions through adjustments in rates of evapotranspiration and albedo. In addition, the shielding of the roofing membrane itself from direct exposure to sunlight and other weathering elements extends the expected life of the roof from an average of 20 years to 40. When accounting for extended longevity and energy cost savings, green roofs are often cheaper to install and maintain than conventional roofs.

For buildings with a low roof-area-to-floor-space ratio, as found in structures more than a few stories tall, the cultivation of vegetation along exterior wall surfaces also can greatly expand their hyrdrologically active envelope. Although it may be difficult to imagine a city populated by creeping vine-laden buildings, for regions that remain well endowed with rainfall, such a low-tech strategy could yield immense returns for a relatively modest investment. The U.S. Environmental Protection Agency, increasingly concerned about the near-term threat of rising temperatures in cities, reports that the integration of vegetation along vertical building surfaces can lower wall temperatures by more than 30°F, translating into significant air-temperature reductions in proximity to the building [15]. The key benefit of such "green walls," as with green roofs, is that they not only expand the hydrologically active area of cities but also do so through conversion of the most highly radiating surfaces – greatly enhancing their thermal benefits.

Such an ecological architecture is already taking root on an experimental basis in large global cities. Perhaps the most well known of these projects is Paris's Musee du Quai Branly, which integrates a 8,600-square-foot green wall composed of 15,000 individual plants into the building's exterior – an installation that is said to draw as many visitors to the museum as the artwork inside. The key innovation on display at the Quai Branly is the use of a nutrient-water-drip irrigation system in place of soil. Developed by Patrick Blanc, a botanist at France's renowned Centre National pour la Recherche Scientifique, the bio-wall system is constructed of a metal and plastic scaffolding attachable to any building exterior, a layer of felt in which the plantings physically take root, and a drip-irrigation system that trickles passively down the wall's surface to a collection tray at the bottom, from which it is then recycled.

The simple system enables a wide range of plant species, including a range of mosses and liverworts that grow naturally on rocky vertical surfaces, to be integrated into almost any building surface, cooling the ambient air as it flows across the exterior.

To date, Blanc has installed more than 150 green walls in more than 25 countries situated in a wide array of climate types. The technology has been found to be durable, with some projects now more than two decades old, and scalable, with increasingly large buildings being draped with bio-walls. Among the most ambitious projects underway is a federal building undergoing renovation in Portland, Oregon, which will incorporate an almost football-field-sized green wall along its 18-story façade – a massive undertaking that will demonstrate the potential for even the largest office buildings to abate their thermal footprints. Much like their green-roof siblings, green walls are no longer novelty architecture but increasingly are viewed as basic infrastructure in a warming world. And, importantly, much like Manhattan's Fifth Street Farm, climate management is only one of an array of benefits associated with building-integrated vegetation, which is often more highly prized for its aesthetics or community-oriented benefits than for its ability to cool the ambient environment and save energy.

However effective green roofs and green walls might be at the building scale, to measurably influence urban-scale climates, such strategies will need to be widely employed in an extensive resurfacing of the built environment. Likewise, the planting of even a million new trees in most large cities will not be sufficient to markedly slow the pace of warming. What is needed more than a short-term greening initiative is a fundamental reengineering of the development process to incorporate energy-balance considerations into all land-use decision making. To this end, the suite of land-development regulations that serve as the DNA for spatial change in cities must be retooled to account for the climate implications of surface modifications. For green roofs to replace asphalt roofs on a large scale, they must not only be permissible but also advantageous under the municipal codes invoked through the issuing of construction permits. These advantages must be made apparent for every resurfacing decision made by public and private landowners. In the form of these codes, numerous levers are available to municipal governments interested in counteracting extreme heat.

The first step is triage. If there is an upside to our present economic maladies, it is the quieting of chainsaws in the formerly hot housing markets of U.S. metropolitan regions. As attested to in Figure 4.1, the rapid and expansive land development in large cities in the past few decades has taken a substantial toll on urban forests. To protect what remains, municipal governments will need to strengthen and vigorously enforce tree-conservation ordinances. Although such ordinances have long been adopted by cities to protect and plan for trees on public property, it is only in recent decades that many large cities have developed policies to safeguard and restore tree canopy on private property undergoing redevelopment. Recognition of the direct implications of deforestation for soil erosion and stormwater runoff has enabled both city and state governments to fend off legal challenges to these provisions by demonstrating a clear linkage between tree preservation and community welfare in the form of clean water and flood control, as well as other environmental services.

In my own city of Atlanta, a tree-conservation ordinance has been adopted with a rigorous underlying standard of "no net tree loss" within the city's boundaries. To meet this standard, developers and other private property owners must file a tree-replacement plan with the city demonstrating how trees lost during the construction process will be replanted elsewhere on site. In addition, applicants for tree-removal permits must pay a sum into the city's tree trust fund based on a formula reflecting the number and diameter size of trees removed. Although these funds provide support for the planting of new trees in strategic areas around the city, such as along roadways, recent inventories on the number of trees felled and the number replanted suggest that Atlanta's no-net-tree-loss standard is not being met. Compared to the more than 15,000 tree removals permitted each year, Atlanta's tree trust fund is supporting the replanting of only about 3,000 new trees annually.

If cities in naturally forested regions are to have any success in slowing an accelerating pace of warming, municipal governments will need to become far more serious about protecting trees. The limited data available on tree protection suggest that although they are woefully deficient, Atlanta's policies remain among the most stringent in the country. However, even if fully enforced, such no-net-loss policies are not designed to increase the extent of the urban forest by a single tree over baseline levels. Shifting from the preservation to the aggressive cultivation of vegetative cover in cities will require altogether new policy tools.

In recognition of the many benefits of green roofs, numerous municipal governments have adopted zoning incentives to encourage their more widespread use. Portland, Oregon, for example, offers developers a "density bonus" for projects incorporating green roofs covering 60% or more of new roofing area. Through this policy, developers are permitted to construct an additional 3 square feet of building floor area per 1 square foot of green roof, enabling the construction of more leasable or saleable built space than would otherwise be permitted under the zoning code. Similar policies are on the books in Chicago, Seattle, and Toronto. For cities in which a stormwater fee is assessed on property owners based on the lot area of impervious surfaces, known as a stormwater utility, green-roof development has been incentivized through a reduction in this fee. Here again, Portland is a national proving ground, allowing discounts on stormwater fees up to 35% in exchange for green-roof installations.

Perhaps the most promising policy innovation related to climate change in cities is the *green area ratio*. A long-standing zoning tool in Germany, green area ratios specify minimum vegetative cover requirements for privately owned property but provide wide flexibility in meeting those cover standards. Functioning much like the Leadership in Energy and Environmental Design (LEED) standards for green-building certification, green area ratio standards identify a menu of greening options from which landowners can choose, including tree planting, green roofs, and green walls. Each strategy is assigned a quantitative score based on its performance across a range of environmental attributes, so that landowners may combine multiple strategies to meet the minimum required point totals.

In 2007, Seattle, Washington, became the first U.S. city to adopt a green area ratio based on the German model – known as the "green factor." At present, the *green factor* specifies a minimum quantitative score that must be met by developers in commercial and multifamily residential land zones, the development classes found to incorporate the most extensive impervious cover. Like other zoning tools, the green-factor requirement must be met at the time of development or redevelopment, when landowners apply for construction permits from the city. As illustrated in Figures 4.4 and 4.5, a wide array of strategies are available to landowners to minimize the ecological impacts of urban development, not all of which require vegetative plantings. Rainwater collection techniques and permeable paving, for example, can be used on site to earn points toward the minimum green-factor requirement. In addition to its relative effectiveness, each strategy is rated as to its projected expense, providing landowners with cost-effective options. For example, green walls, the most expensive strategy, are assigned a lower score than rain gardens, among the least costly strategies.

That said, a significant expansion of green space on urban parcels will impose nontrivial costs on landowners, particularly the owners of single-family homes, which often account for the majority of developed

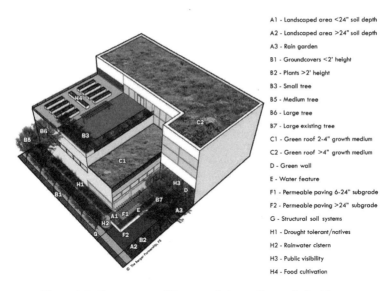

A1 - Landscaped area <24" soil depth
A2 - Landscaped area >24" soil depth
A3 - Rain garden
B1 - Groundcovers <2' height
B2 - Plants >2' height
B3 - Small tree
B5 - Medium tree
B6 - Large tree
B7 - Large existing tree
C1 - Green roof 2-4" growth medium
C2 - Green roof >4" growth medium
D - Green wall
E - Water feature
F1 - Permeable paving 6-24" subgrade
F2 - Permeable paving >24" subgrade
G - Structural soil systems
H1 - Drought tolerant/natives
H2 - Rainwater cistern
H3 - Public visibility
H4 - Food cultivation

Figure 4.4   Components of the green-factor ordinance in Seattle, Washington. *Source: Functional Landscapes: Assessing Elements of Seattle Green Factor*, The Berger Partnership PS, 2008.

land in urban areas. Therefore, municipal governments will need to couple such policies with financial incentives to offset the initial investment in potentially cost-prohibitive projects such as green roofs or permeable paving. Yet, so great have become the public costs of managing environmental problems associated with urban imperviousness that many cities are finding it cost effective to directly underwrite some of the expense associated with greening initiatives on private property. Toronto, for example, has established a program through which the city will cover up to $50 per square meter of green roof retrofitted onto existing industrial buildings – enough to cover at least 20% of most green-roof projects – up to a total cost of $100,000. Likewise, Chicago, Cincinnati, New York, Philadelphia, and Portland offer grant programs or tax credits to support the installation of green roofs and green walls.

Preliminary data on the new program in Seattle suggests that it is having a direct effect not only on the parcel area devoted to impact mitigation but also on the types of vegetation incorporated into the site. Although permit applications in the 16 months following the adoption of the green-factor requirement showed that conventional landscaping, such as lawn area, was the most prevalent strategy employed by area of coverage, green roofs were the second most prevalent – accounting for almost 20% of the total new green area developed under the

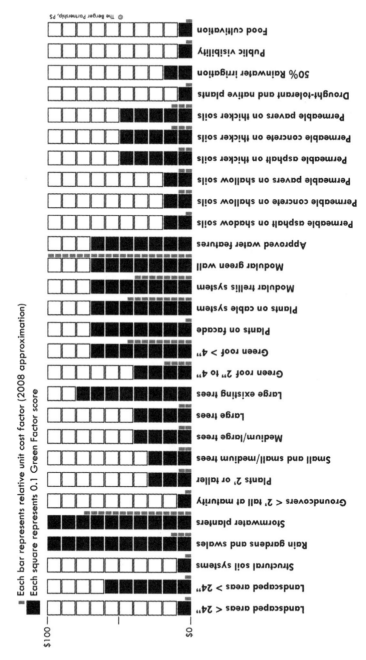

Figure 4.5   Relative weighting and cost estimates for green-factor strategies in Seattle. *Source: Functional Landscapes: Assessing Elements of Seattle Green Factor*, The Berger Partnership PS, 2008.

program [16]. Such a shift toward the construction of green roofs is encouraging because this strategy carries the greatest potential of any approach to reduce both land-surface temperatures and building-energy demands. The costs thus far have been far from prohibitive: developers of commercial projects report the cost of meeting the green-factor standard to be 0.4% of total project costs.

Green area ratios provide what is likely to be the single most effective policy tool to combat extreme heat in cities because of their leveraging of market-driven redevelopment activities to bring about more climate-responsive built forms. In contrast to public tree-planting campaigns, the reliance of green zoning tools on private-sector development and redevelopment activities expands climate mitigation into all zones of the city and, if broadly applied, times its implementation with routine maintenance such as the periodic resurfacing of a roof, driveway, or parking lot. It also has the effect of shifting the cost of impact mitigation to the impact generator – the property owner.

More challenging perhaps than the expense of incorporating greater vegetative cover into urban environments will be the provision of sufficient water resources to support extensive plant life. As the global greenhouse effect shifts precipitation patterns over time, both the frequency and intensity of rainfall are likely to present problems for establishing and maintaining trees and other vegetation. In most regions of the world, heat- and drought-tolerant species of trees and other vegetation will be required. As traditional climate patterns continue to shift, urban arborists will need to consider species selections outside of the collection of trees and plants considered native to a particular region.

Where the strategic selection of species is not sufficient to ensure the survival of new trees, technological intervention may be required as well. Use of an innovative water-saving device created to enable tree growth in desert environments is likely to become increasingly commonplace in cities. Known as the Groasis Waterboxx, the device is affixed around the base of a newly planted tree to capture condensation from the atmosphere and regulate its delivery to the plant. A simple box-like device in design, the key innovation of the Waterboxx is its inhibition of evaporation. Once collected, both condensation and, when available, rainfall can remain in the device for extended periods without evaporating, even in hot and arid environments. Tests of the tire-sized device in the deserts of Morocco have demonstrated its effectiveness, with about 90% of trees planted with the box developing roots of sufficient length to reach underlying soil moisture, compared to only 10% of trees planted without it, even when watered daily [17].

Little more in appearance than a fancy bucket, the Waterboxx may be the most significant climate-related invention in a generation, and its potential has not gone unnoticed. In 2010, the Waterboxx – alongside the venerable iPad – was awarded a best-of-the-year technology prize from *Popular Science* magazine. Through enabling the establishment of trees and other plants in regions otherwise inhospitable to shallow-root vegetation, the simple device enhances the possibility of dramatically greening urban environments in which the cultivation of trees is becoming increasingly challenging. In doing so, the Waterboxx demonstrates a key principle of urban-climate management: cities themselves must become more efficient devices for the collection, retention, and utilization of water – an imperative attained more easily in some regions than in others.

It is sometimes too hot to fly in Phoenix. The problem arises not from the potential for airport systems failures, although power-system shutdowns and the inability of ground crews to service the planes under conditions of extreme heat have certainly closed airports in the past. The risk of flying in Phoenix on very hot days is caused by something more elemental: the low density of the air. Under the superheated conditions that have been experienced in past heat waves, reflected by temperatures in excess of 120°F, the density of the air can drop so low as to impede the ability of aircraft to sustain sufficient lift during takeoff and landings. As the occurrence of such extremely hot days grows more common with climate change, Phoenix and other desert cities will likely need to redesign their transportation systems in several ways, including developing longer runways to enable large jets to achieve liftoff during periods of extreme heat.

Airport operations will not present the only challenge unique to cities in hot and arid climates. More pressing will be the need to manage extreme heat in heavily populated areas lacking access to ample water supplies. For although water-saving devices such as the Waterboxx hold great promise for the cultivation of plants ultimately able to access subterranean moisture, there is little potential for such inventions to support building-integrated vegetation strategies such as green roofs and green walls. In these environments, urban-climate management will need to emphasize an altogether different physical mechanism of heat mitigation: surface reflection.

Enhanced albedo is the technique the desert uses to cope with high levels of solar exposure. Unshielded by a dense canopy of vegetation and lacking sufficient soil moisture to offset heat gain through

evapotranspiration, the lightly hued soils and sands of desert environments reflect away a significant percentage of incoming solar radiation
before it is absorbed and transferred to thermal radiation and sensible
heat. Although high levels of albedo alone will not bring about mild
afternoon temperatures in desert climates, enhanced reflectivity has
long been used in hot and arid cities to moderate temperatures. The
use of whitewashed coatings throughout the Mediterranean and Middle
East, for example, combined with other passive cooling strategies embodied in courtyard housing styles, can significantly reduce both indoor
and ambient temperatures, rendering such climates habitable without
mechanical air conditioning.

Recognition of the potential to measurably cool cities through the
application of highly reflective coatings to roofing surfaces and streets
has led to the development of new product lines in the roofing and paving
industries. For roofing surfaces concealed from ground view, such as
atop a flat industrial building, very high-albedo, white surface coatings
can be applied to reflect away a substantial percentage of incoming
solar radiation. Industry analyses of these materials have found that the
surface temperature of the roofing materials themselves can be reduced
by as much as 50°F during periods of intense solar gain.

To explore the extent to which cool-roofing materials could reduce
temperatures not only within the treated buildings themselves but also
throughout the ambient urban environment, scientists at the Lawrence
Berkeley Labs and elsewhere have modeled extensive albedo enhancement strategies. The key question concerns the extent to which the
average reflectivity for a full city can be increased through cool-roofing
treatments. Measured on a scale of 0 to 1, average surface albedos in U.S.
cities tend to range from 0.10 to 0.20, much lower than the albedo range
of 0.30 to 0.45 found in the whitewashed cities of North Africa [18]. In
densely settled districts such as Manhattan, in which rooftops account
for about 40% of the total area, the potential to raise average albedos is
great, but all cities can enhance their reflectivity.

In Los Angeles, for example, the conversion of all flat or low sloping roofs to highly reflective materials could lower average summer afternoon temperatures by an amount equivalent to the planting of 11 million
trees: almost 3°F. Given that these strategies have been found to be mostly
additive in their effects, the combination of aggressive tree planting with
extensive roof conversions to high-albedo materials throughout the Los
Angeles basin has been shown through computer simulations to reduce
summer afternoon temperatures by more than 5°F – enough to fully offset the region's typical heat island [4]. This finding is critically important

in two respects. First, it is suggestive of the potential to substantially slow ongoing warming trends through a combination of strategies that could feasibly be carried out during the period of a decade or two. Second, the long-term economic savings in the form of reduced energy consumption and avoided air-pollution management costs would exceed the initial investment. Taken together, these observations point to an unexpected outcome: the most effective means of slowing the pace of warming in Los Angeles in the next few decades comes not with an exorbitant price tag but instead with a profit.

A key advantage of albedo enhancement over other urban-climate-management strategies is its relatively low cost. Cool-roofing treatments can be applied to low-sloping roofs for a cost premium of between $0.05 to $0.10 per square foot, raising the cost of a 1,000-square-foot roofing project by about $100. Balanced against this low initial cost are annual energy savings estimated by the U.S. EPA to be about $0.50 per square foot, an estimate accounting for potentially greater heating costs in winter [15]. Therefore, the payback period on most cool-roofing projects is less than a year. Also advantageous is the immediacy of beneficial returns from cool-roofing strategies. In contrast to tree planting and other vegetative programs, through which maximum cooling benefits are not realized until plants reach maturity, high-albedo coatings yield maximum benefits at the time of installation, with benefits diminishing somewhat thereafter with weathering.

What may be a greater obstacle to the widespread adoption of highly reflective roofs than the installation costs is an aesthetic preference for darkly hued roofing materials. A cultural legacy of Northern European immigrants to North America, the darkly hued slates and other roofing materials in their home countries serve an important climatic purpose in maximizing solar absorption in a predominantly cold climate. However, the extensive use of black asphalt shingle across lower latitude cities makes very little economic or environmental sense in a warming world. With an albedo close to zero, black roofing materials alone can increase building cooling costs by 10% to 20%, resulting in higher ambient temperatures and copious amounts of excess greenhouse gas emissions. Replacing such materials with more highly reflective roofs at the time of routine maintenance could reduce summer afternoon temperatures in many large cities by several degrees. Consistent with findings from Los Angeles, a study of 10 of the most populous U.S. cities suggests that a combination of albedo enhancement and greening strategies could reduce heat-island formation by between about 25% and 75% [19].

Even a distaste for lightly hued roofing materials is today no real impediment to offsetting heat gain at the building envelope. A greater understanding of the thermal characteristics of various materials has enabled the engineering of more highly reflective roofing shingles across a wide palate of colors. The development of roofing materials that quickly radiate away absorbed heat energy, a property referred to as surface emissivity, enables even darkly hued shingles to significantly offset heat gain. In combination, the modification of surface albedo and emissivity has enabled an increase in the effective albedo of black roofing shingles from 0.04 to more than 0.40 – a feat of engineering that could yield substantial cooling dividends to cities if strongly incentivized through government policies.

The beginnings of such a policy framework are already on the books in the United States, which is quickly emerging as the global leader in both high-albedo construction materials and policy innovations. As with green area ratios, the most effective policies are municipal regulatory codes governing construction and building improvements. On this front, Austin, Texas, has adopted the most ambitious policy in the United States, requiring low-sloped roofs to have a minimum albedo of 0.70 and all other roofs to meet a minimum albedo standard of 0.35 – exponentially more reflective than a standard black-shingle roof. Other cities mandating minimum reflectance values for all or some buildings are Chicago, Dallas, Houston, and New York. At present, nine states have mandatory minimum-reflectance requirements for various classes of buildings, ranging from state-owned structures to all commercial and residential buildings. In addition, a large number of cities and states, as well as the federal government, offer rebates and/or loan programs to offset the generally modest cost premium of cool-roof installations.

High-albedo roofing materials offer a very promising strategy for slowing the pace of urban warming over the near term, but the potential for incorporating highly reflective pigments into street paving and vertical building surfaces is more limited in a few respects. The most intuitive disadvantage is glare. Because the bulk of the shortwave solar radiation reflected by high-albedo surfaces is visible light, such reflection at ground level can create visibility problems for drivers and pedestrians. Related to this problem is the issue of radiant exposure. When absorbed by the human body, reflected radiation raises skin temperatures, creating the very problem that high-albedo strategies are designed to avoid. Nonetheless, a moderate increase in the reflectivity of very low-albedo paving materials, such as black asphalt, would offset the thermal impacts of a larger fraction of the urban footprint than roof area alone.

In combination, the albedo enhancement and vegetative strategies considered in this chapter can be understood to physically screen urban environments from the accumulation of heat through radiant reflection, shading, and evaporative cooling and thus may be characterized as a form of urban "sunscreening." Although all cities can benefit from sunscreening techniques, the extent to which vegetation or albedo enhancement should be pursued in any particular context is dependent on regional climatic conditions. In tropical to temperate zones projected to retain sufficient annual precipitation over time, tree planting and building-integrated vegetation are likely to offer the most effective climate-management strategies, combined with albedo enhancement where dense development limits planting opportunities. In the hot and arid climates of the U.S. Southwest and other desert environments, albedo enhancement should be emphasized over vegetative strategies to limit additional stresses on diminishing water resources. In either instance, the built environment of cities must be made to more closely mimic the natural environment of which they are a part – increasing the climate resilience of both in the process.

It is important to emphasize that the sunscreening of urban cores alone may not be sufficient to manage warming trends driven by landscape change across a much larger region. As revealed by elevated warming trends in cities experiencing extensive metropolitan deforestation, discussed in Chapter 3, land-use change outside of city centers can play a role in the frequency of extreme heat within the urban core itself. To counteract this trend, it is important that forest and other natural land covers in proximity to large cities be protected and regenerated to create a metropolitan greenbelt. Referred to herein as urban "greenbelting," the regeneration of forest and other land covers conducive to regional moisture retention and evaporative cooling, such as irrigated farmland, is a key component of a regionally comprehensive approach to urban-climate-change management.

A significant policy challenge related to greenbelting-strategies concerns the ability of urban governments to influence land-development patterns beyond their jurisdictions. In contrast to vegetation and albedo enhancement within the city itself, greenbelt protection and reestablishment require well-coordinated land-use management across a metropolitan region. Where state and/or national governments have empowered metropolitan areas to coordinate land-use planning across local jurisdictions, cities may be able to establish "urban growth boundaries" to protect natural areas and farmland from the expansion of developed land. However, where such planning tools are not commonly found, as throughout most of the United States, state or provincial governments may need to act directly to establish climate-protection zones around large cities. For cities lacking regional planning authority, the protection of greenbelt areas through the purchase of land or development rights offers a more direct, albeit far more costly, approach to regional climate management.

In addition to sunscreening and greenbelting strategies, a final approach to urban-climate management entails reductions in the direct emission of waste heat from industrial processes, buildings, and transportation systems. Such "carbon-cooling" strategies, as they will be characterized, provide a critical linkage between regional- and global-scale climate-management activities and are the focus of the chapter's concluding section.

If your city feels hotter after a day at the office than after a day at the pool, it probably is – and the reason has little to do with your clothing. Perhaps the least well-understood driver of climate change in cities is waste heat. Emitted from smokestacks, tailpipes, and the human body, among numerous other sources, waste heat contributes to warming in cities brought about by the increased radiant and sensible heat fluxes attributable to land-surface changes. In the most basic terms, the heat produced through fossil-fuel combustion augments the warming of present-day sunlight with ancient sunlight, long entombed in the coal, oil, and natural gas ultimately burned in urban areas. Naturally, the more of this energy that is combusted in a city, the more heat is injected into the atmosphere – which explains, at least in part, why cities may be hotter during the week than on the weekend.

To confirm this theory, a team of researchers from Portland State University made use of a heat-island measurement technique first employed in the 1960s, before the launching of satellites with thermal sensors: they affixed a thermometer to the roof of a car. Once jury-rigged in this fashion, the car provides a mobile sensor that is driven in transects from the city center to the periphery, enabling measurement of the contours of a heat island throughout an urbanized area. In comparing these measurements to rural weather-station observations, the researchers were able to map heat-island intensity throughout Portland. They were also able to measure the variation in heat-island intensity at different times through the collection of data on both week and weekend days.

Consistent with other large cities, temperatures in the most intensely developed districts of Portland were found to be significantly higher than in rural areas, with a heat-island intensity of 9°F. Heavily canopied areas within the city, such as densely forested parks, were found to be as much as 18°F cooler than the most built-up districts, attesting to the strong potential for tree planting and other vegetative approaches to mitigate the extent of warming in Portland. Most important, the difference between urban and rural temperatures on the weekends was found to be consistently lower than on weekdays, with heat-island intensity in the downtown district between 2 and 4°F lower on weekend days. This difference, the researchers believe, is attributable to the greatly reduced number of vehicles on the roads during the weekend in downtown Portland, as well as to reduced industrial activities and the presence of fewer human bodies in general [20].

The potential for waste-heat emissions to measurably raise ambient temperatures has been recognized for decades. Simple energy balance calculations reveal the importance of waste-heat emissions in high-density cities. In Manhattan, for example, the total quantity of waste heat released from buildings, industry, and vehicles exceeds the quantity of absorbed solar energy released as sensible heat from the Earth's surface. In most North American cities, however, the ratio of combusted heat to surface-emitted heat is much lower, particularly in summer, with waste heat accounting for less than a fifth of the total heat-energy burden to be managed in lower-density cities such as Los Angeles [21]. Nonetheless, waste heat is an important driver of rising temperatures in all cities, presenting an additional target for urban-climate management.

The most significant source of human-emitted or anthropogenic heat in cities is vehicle traffic. Based on traffic densities in large U.S. cities, vehicles are estimated to account for about 50% to 60% of the total waste-heat burden during the summer months, when heat islands present the greatest threat to human health. Most of the remaining heat emissions are generated by industry or building air-conditioning systems, which mechanically transfer heat from building interiors to the ambient air. Assuming average population densities of about 10,000 per square kilometer during the workday, heat produced through human metabolism typically accounts for 2% to 3% of the total summer waste-heat load in large U.S. cities, a small but measurable contribution to heat-island formation[1] [22].

Anthropogenic heat emissions have been found to directly elevate urban temperatures by between about 3 and 5°F in the downtown cores

---

[1] Sailor & Lu [22] find the human metabolic heat flux to account for less than 5% of total anthropogenic heating in most U.S. cities, with the contribution believed to be higher in countries with lower energy consumption per capita.

of large cities, accounting for a lower magnitude of warming in lower-density residential districts [18]. The significance of waste heat to urban warming suggests a number of strategies that can be combined with changes to physical surface properties to slow warming – strategies that I refer to collectively as "carbon cooling" due to their potential to reduce both waste heat and greenhouse emissions. Perhaps the most evident of these strategies is a greater reliance on urban transit systems to reduce vehicle use during periods of peak congestion, which, in the instance of the weekday afternoon commute, typically coincide with peak levels of heat-island intensity. The extent to which a particular mode of travel contributes to heat-island formation is directly related to its energy intensity, a measure of the energy consumed per passenger per unit of distance traveled. In the United States, the energy intensities of urban rail transit systems are about a third lower than the energy consumed by cars and light trucks [23], highlighting the clear potential for expanded mass-transit usage to directly counteract rising temperatures in cities.

To be effective as a heat-management strategy, personal vehicle use would need to decline substantially in favor of mass-transit usage. Although such an ambitious goal has long been out of reach in all but the most populous U.S. cities, declining global oil reserves may bring about such changes in the near term, entirely independent of climate considerations. Yet, investment in transit systems, particularly rail systems, is a sound climate-management technique in its own regard. Electricity-driven rail systems not only produce far less waste heat per passenger than private vehicles, the production of this heat also is centralized at remote power stations, typically situated outside the most heavily populated districts of a metropolitan region. The centralization of this waste heat enables its use for other purposes, further diminishing the total quantity of energy consumed and waste heat produced. More important to a comprehensive climate-management program, however, is the extensive availability of urban transit as a prerequisite to the population densities needed to shrink metropolitan footprints and better enable the sunscreening and greenbelting strategies already discussed.

Urban densification must be the centerpiece of any serious climate-management program undertaken at the metropolitan scale. Fundamentally what this means is that regional economic-development strategies rooted in low-density, sprawling housing development at the urban periphery are wholly untenable in a warming world. Although such assertions of the obvious may not curry favor among land developers, neither the rate at which cities are presently warming nor the rate at which global reserves of conventional oil are declining bodes well for the continuing horizontal spread of cities. Rather, it is quite likely that recently sprouted housing developments, many only partially constructed in the wake of the great recession, are in the initial stages of a Detroit-like reversion to the pastureland and forests only recently cleared for their construction. As urban temperatures routinely climb to dangerous levels, the demolition of peripheral, low-density development to enable a restoration of heat-abating forestland may increasingly be viewed as a viable climate-management approach.

Although higher-density urban-development patterns are generally assumed to be conducive to heat-island formation, just the opposite is true when heat production is measured at the metropolitan scale. Because sprawling development tends to result in both more land clearance and more impervious cover per capita, the quantity of excess heat generated per person in lower-density settings is substantially more than in higher-density urban cores. In a study of excess heat production across more than 100,000 single-family residential parcels in the Atlanta region, we found the quantity of excess heat emissions per lot to increase by more than 30% with each additional quarter-acre of parcel area [24]. Thus, a family of four living on an acre-sized parcel in Atlanta generates more than twice the excess heat energy of a comparable family living on an eighth-acre parcel, a size consistent with many urban, single-family neighborhoods. Because convective wind patterns generated by heat islands tend to channel heat toward the downtown district, excessive temperatures experienced in the urban core are a product of development patterns found throughout a metropolitan region.

Rather than exacerbating heat islands, higher-density development, when balanced with diminished rates of development outside of city centers, enables heat-island mitigation through amassing the resources needed for sunscreening strategies. In this sense, the per-acre expense of installing green or white roofs is far more manageable when shared by eight families than when borne by one, particularly if these families live in the same building. Therefore, well-planned and extensive transit systems that enable urban residents to travel through cities at a lower cost and in less time than personal vehicles, as found in many European and Asian cities today as well as in a few U.S. standouts, also facilitate climate-management strategies that rely on minimum levels of population density to become economically viable. Coupled with the potential for these systems to reduce waste-heat emissions from vehicle engines, investments in urban transit systems that facilitate higher-density modes of living may represent one of the most effective means of managing climate change in urban areas.

The reduction of waste heat from buildings, the second most prevalent source of waste heat in cities, can be achieved in part through

the sunscreening strategies discussed throughout this chapter. Enhance-
ments in evapotranspiration and reflectivity brought about through
building-integrated vegetation and high-albedo roofing materials can
greatly reduce air conditioning, limiting the transfer of heat to the
ambient air. In larger-scale heat production, as associated with industrial processes, the utilization of waste heat for a secondary purpose can
further offset waste-heat emissions. Through a "combined-cycle" power-
plant design, for example, the exhaust heat from gas-fired turbines is
captured and used for a secondary purpose, such as the generation of
steam to drive a second turbine and produce additional electricity. In
some power-plant designs, the secondary steam can be used to drive
chiller systems to cool buildings. In both instances, the use of waste heat
for a secondary purpose offsets the total quantity of fuel burned and
waste heat emitted to the atmosphere.

Of course, any strategy that seeks to improve energy efficiency carries with it the co-benefit of carbon cooling. An inviolable implication
of the second law of thermodynamics is that waste heat results from all
conversions or uses of energy, with the quantity of heat produced deter-
mined by the efficiency of the process at hand. Therefore, any technique
that enhances the energy efficiency of a building or the operation of a
vehicle engine also reduces waste-heat emissions. It is with respect to this
observation that one of the most compelling reasons for aggressively pursuing land-based mitigation in cities is made plain: almost any strategy
imaginable for slowing the rate of warming in cities is also an effective
strategy for reducing greenhouse gas emissions. Whether in the form
of climate-responsive building design, less energy-intensive transportation systems, or more efficient power-generation processes, strategies
effective in reducing ambient temperatures in cities are almost always
effective in reducing energy consumption for cooling or power generation – the principal drivers of greenhouse gas emissions.

In this sense, climate-change mitigation at the urban scale is wholly
compatible with climate-change mitigation at the global scale. Yet, the
incentives for locally and globally responsive strategies are quite different, a key but seldom observed fact that opens the door to entirely new
policy approaches to climate management. Home to more than half of
the global population and responsible for the majority of global greenhouse gas emissions, cities are the domain in which both the principal
causes and human effects of climate change are playing out. But, as
of yet, cities have enjoyed no unique status in international climate-
policy frameworks. Fundamental to this oversight is a failure on the
part of most climate scientists and policy makers to recognize the profound significance of land use, the other half of the global greenhouse

equation. As I argue in the book's final chapter, aggressive management of both the atmospheric and the land-surface drivers of climate change is our only viable option for slowing the rapid pace of warming already underway.

It has been my intent in this chapter to examine the policy options available to cities to measurably slow the pace of warming and the incidence of extreme heat in the near to medium term. A review of the principal mechanisms for mitigating climate change in cities shows that municipal governments are empowered to play a major role in managing their own climate futures with the array of policy tools and planning expertise already at hand. Although the extensive resurfacing of urban landscapes needed to enhance climate resilience will require significant public- and private-sector investment over the coming decades, much if not all of this investment can be recouped in the form of long-term energy savings and restored environmental services.

Because many of these changes will be necessitated by rising energy prices, the principal question posed to urban governments concerns not what to do but how soon to start. The answer to this question will depend very much on the degree to which national and international climate-management frameworks recognize land-based mitigation at the local and regional scales. With international climate negotiations at a standstill, there is much evidence to suggest that a restructuring of global frameworks to respond to a wider set of management actions is needed to resolve a long-standing impasse between developed and developing nations, enabling both to exploit respective comparative advantages in slowing the global greenhouse effect. What such a restructuring might entail and its implications for cities, in particular, are my focus in the final chapter.

5

Leveraging Canopy for Carbon

That all Dutch children must learn to swim by the age of six is a national mandate that probably makes sense. The Netherlands, of course, is a country of water – ocean, inland seas, rivers, and canals. A legacy of all those iconic windmills, vast regions of the Netherlands have been reclaimed from inland water bodies through massive water-pumping projects initiated in the 15th century and continuing ever since. Today, a quarter of the country's land lies below sea level, with more than 20% of the Dutch population residing in these areas. Of the remaining land area, most sits at only a few feet above sea level. So the likelihood of encountering water as a swimmer, even at a young age, seems great. What is perhaps most remarkable about this requirement, however, and most revealing about the Dutch national character, is not that all children must demonstrate proficiency in the water at a young age but rather that they must do so fully clothed. The Dutch are not preparing for a day at the beach; the Dutch are preparing for an emergency.

They have good reason to do so. At its lowest points, the Netherlands sits at 23 feet below sea level, exceeding the lowest elevations in New Orleans, for example, by about 16 feet. Such a pronounced change in elevation contributes not only to the depth of flood waters when levee systems fail but also to the rate at which flood waters rise. In past flooding events, the most catastrophic in February 1953, flood waters surged into low-lying areas without warning, yielding in a matter of hours a death toll equivalent to that of Hurricane Katrina. In the aftermath of the 1953 event, the Dutch undertook the construction of a massive flood-management system that would require almost a half-century to complete. Consisting today of more than 8,000 miles of levees, dikes, and sea barriers, the water-management system now protecting the Netherlands from rising seas is regarded by the U.S. Army Corps of Engineers as one of the seven wonders of the engineering world. But it will not safeguard the Dutch from climate change.

In recognition of this fact, the Dutch are raising and fortifying their sea defenses in some places and dismantling them in others. Drawing on a half-millennium of experience in managing water, the Dutch are fully cognizant of the limitations of hard infrastructure in the face of a rising ocean. Therefore, current long-range plans call for floodwaters to be channeled into some low-lying areas during storm events to relieve pressure on other parts of the water-management infrastructure, particularly around the most heavily populated regions. These plans are more than a tentative policy goal to be fulfilled at a future date: the Netherlands government is at this very moment purchasing land, resettling residents, and demolishing buildings in designated overspill zones. It does so not to comply with any mandates or incentives under international climate-change agreements but rather because rising sea levels pose a direct and existential threat to the country and its population. Widely lauded for their ingenuity in managing rising seas, an additional insight to be drawn from the Dutch is this: the most consequential actions being taken by nations to prepare for climate change almost always fall outside the parameters of international climate agreements.

A corollary to this observation is that the most consequential actions being taken by nations to prepare for climate change almost always involve land-use change. Whether such programs entail the construction of levees, the planting of trees, or the development of new reservoirs, the climate-change–related actions that are yielding measurable benefits in the present period are often associated with physical changes to local environments that may or may not have a direct influence on emissions reduction. What this basic observation suggests about climate-management policy is a seemingly intuitive but rarely observed truism central to global climate governance: *to be effective, actions taken in the interest of the global community must also be central to the interests of the local community.*

It is my intent in this final chapter to explore the implications of global climate policy for climate-change management at the urban scale. In doing so, I offer a broad critique of the failure of international agreements to sufficiently incorporate land use into the global framework for climate-change management. As presently structured, the international climate policy framework addresses only one of the two physical drivers of climate change: the emission of greenhouse gases. A failure to counteract the second set of climate-change agents, land-surface characteristics related to albedo and the surface-energy balance, has not only produced a governing regime that overlooks what are often the principal drivers of climate change in the places we live but has also created a framework that is not particularly effective in achieving its singular objective of emissions control.

The reason for this perverse outcome is that an "emissions-only" orientation to climate management ultimately decouples local interests from global interests and, as a result, obscures the community-scale benefits of participation in the global climate policy regime. A better approach would seek to marry the priorities of local communities, which are more strongly rooted in climate adaptation, to long-term global objectives, which are more strongly rooted in emissions controls. To understand how the international policy framework might be restructured to address both the global and local dimensions of climate change, this chapter explores three basic questions. First, are land-use strategies an effective means of global carbon management? Second, to what extent are land-use strategies being pursued under the international climate policy framework? Third, how can the international policy framework be restructured to more effectively employ land use as a tool for climate management, particularly at the scale of cities?

Before addressing these questions, it is important to explore the basic policy architecture of international climate agreements and to consider how these agreements compound the challenge of managing climate change at the scale of cities and regions.

The need for a restructured approach to global climate management was made plain at the Copenhagen Climate Conference in December 2009. Billed by its promoters as the "Hopenhagen" accords, the principal goal of the global summit was the hammering out of an international agreement to succeed the governing Kyoto Protocol, set to expire in December 2012. If the conference organizers overplayed their hand by hailing the impending adoption of "Mother Nature's bailout plan," high optimism in advance of the conference had seemed warranted for several reasons. A new American president had signaled his support for a strong agreement; the U.S. Congress was debating credible climate-related bills for the first time in a decade; and China, now the world's leading emitter of greenhouse gases, had in the lead-up to the conference established meaningful targets for a unilateral reduction in emissions. Most important, direct evidence of accelerating warming abounded, with each of the 10 hottest years ever measured occurring since the signing of the last major global accord in 1997, and a growing list of climate-related events ranging from the European heat wave to Hurricane Katrina to drought-induced crop failures across the planet.

These events notwithstanding, no new emissions treaty would be formally adopted by the participant nations in Copenhagen. In fact, as I write in 2011, no new policy framework has yet been established to replace or extend the soon-to-expire Kyoto Protocol, the only global agreement in place to limit emissions of greenhouse gases. So resounding was the failure of the Copenhagen talks that more than a few highly respected climate-policy experts would conclude that our last, best chance for averting substantial societal impacts had passed and that national governments would be wise to focus their resources on local adaptation strategies. Yet, rather than abandon global strategies for climate management altogether, there are reasons to suspect that it is the limited policy orientation of the current international framework, rather than the scale of governance, that most inhibits the adoption of new binding agreements. To understand why this may be the case, it is first useful to consider how the present policy landscape has come about.

The Kyoto Protocol is actually the second of two major international agreements that together constitute the basic framework for global climate management. The first of these, the United Nations Framework Convention on Climate Change (UNFCCC), was agreed to by participants in the United Nations Conference on Environment and Development in 1992, known informally as the Earth Summit. The issue of climate change had emerged on the global radar first through the work of James Hansen and a handful of other prominent climate scientists in the 1970s and 1980s, and then more formally through the work of the Intergovernmental Panel on Climate Change (IPCC), established by the United Nations Environment Programme and the World Meteorological Organization in 1988 to develop consensus scientific assessments on the issue. In the wake of the IPCC's first assessment report, issued in 1990, the Earth Summit was convened to develop a global response to the increasingly robust science on human-driven climate change.

Although the UNFCCC imposed no legally binding commitments on participating nations, this initial agreement in many respects has shaped global climate policy in more profound ways than the binding treaties that have followed. Most important, it is in this initial treaty that the basic phenomenon of climate change would be formally defined, and it is in this governing definition that we find the origins of the framework's most significant limitations. As defined by the Framework Convention of 1992, climate change constitutes a change of climate which is attributed directly or indirectly to human activity that alters the composition of the global atmosphere and which is in addition to natural climate variability observed over comparable time periods [1].

With respect to the policy architecture that would follow, the pivotal phrase in this definition is "composition of the global atmosphere," a phrase limiting the scope of climate change to emissions-related phenomena only. What is noteworthy about this definition is that it fails to account for the nongaseous drivers of climate change, drivers that influence the magnitude of longwave radiation or sensible heat emitted from the Earth's surface. These latter drivers do not effect a change in the composition of the atmosphere (unless we are to interpret "composition" as encompassing both gaseous and radiative atmospheric elements, an interpretation nowhere in evidence in subsequent management programs) but, as argued throughout the course of this book, nonetheless constitute major agents of human-driven climate change, particularly at the scale of cities.

What is perhaps most surprising about the Framework Convention definition of climate change is that it departs sharply from the definition adopted by the IPCC. In contrast to the UNFCCC definition, the IPCC construes climate change more broadly, defining it as

a change in the state of the climate that can be identified by changes that persist for an extended period, usually decades or longer. It refers to any change in climate over time, whether due to natural variability or as a result of human activity [2].

From a scientific perspective, climate change results from both natural and human drivers and, most important, the types of human activities that may influence climate change are not limited to those affecting only atmospheric composition. The IPCC is fully cognizant of the role of human-driven land-use change in climate phenomena. As stated in its third assessment report, "climate change may be due to natural internal processes or external forcings, or to persistent anthropogenic changes in the composition of the atmosphere *or in land use*" [3, emphasis added]. As stated previously, the more narrow definition of the Framework Convention, which provides the basis for global climate policy, does not recognize land-use change (or any nongaseous agent) as a human climate-change agent that may be addressed through a policy response.

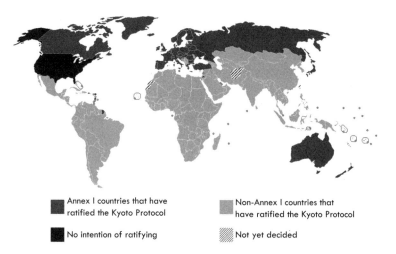

Figure 5.1   National status of Kyoto Protocol ratification. Annex I countries are those that agreed to binding emissions-reductions obligations between 2008 and 2012. *Source*: Adapted from http: //en.wikipedia.org/wiki/File:Kyoto_Protocol_participation_map_2010.png.

Informed by this emissions-only orientation to climate change, the UNFCCC calls for the adoption of legally binding treaties, known formally as "protocols," to establish annual emissions quotas for participating nations. The Kyoto Protocol is the first such treaty and is designed to reduce aggregate emissions of greenhouse gases to an average of 5% below 1990 levels by 2012. Importantly, not all nations that have ratified the protocol are required to meet these emissions targets: developing nations are exempt from binding emissions reductions. As a result, only a relatively small number of countries – 40 of the 194 countries that ratified the treaty – are required to meet emissions targets. These countries are shown in Figure 5.1.

Moreover, the world's largest emitter of greenhouse gases when the agreement was put in place, the United States, is one of only two sovereign nations worldwide (the other being, not unremarkably, Afghanistan) to have failed to ratify or observe the treaty. As a result, given that developing economies represent more than 50% of global greenhouse emissions and the United States today represents about 18%,[1]

---

[1] These emissions trends are for 2005, the most recent year for which complete global data is available, and reflect emissions from both fuel combustion and land-use activities (World Resources Institute, Climate Analysis Indicators Tool, cait.wri.org).

the Kyoto Protocol addresses the emissions activities of countries responsible for only about one-third of annual global emissions. In response to
this shortcoming, organizers of the Copenhagen conference had hoped
to develop a successor agreement that could achieve more substantive
reductions through the participation of the United States and of rapidly
industrializing countries, such as China and India.

Each of these parties – the United States and the most rapidly
industrializing developing countries – has clearly articulated rationales
for opposing present formulations of a new global treaty. The most industrialized of the developing nations are unwilling to accept binding limitations on emissions if such reductions threaten the pace of their own
economic growth. Although the pace and nature of this growth – often
powered by the most carbon-intensive fossil fuels – are accelerating
the accumulation of greenhouse gases in the atmosphere and hasten-
ing adverse impacts in these very countries, rapid development is also
viewed as a key variable in managing the negative impacts of climate
change. Development of extensive water-delivery systems, coastal pro-
tections, and an expansion of industrial agricultural systems engineered
for greater climate resilience will require vast economic and energy
resources in the coming decades. Combined with that concern for adaptive capacity is a strongly held position that it is the global North, with
historical levels of emissions per capita much greater than those of the
global South, that is more responsible for human-driven climate change.
In deference to this position, developing nations have not been required
to accept binding emissions limits under the Kyoto framework.

It is precisely this emissions exemption that dissuaded the U.S.
Congress from ratifying the Kyoto Protocol. Because U.S. industries are
increasingly competing with industry in developing nations, many view
the imposition of binding emissions reductions on one party and not
the other as placing the United States at a distinct competitive disadvantage, with negative implications for economic growth. Proponents of this
position argue further that any treaty failing to address the emissions of
developing economies, now accounting for more than 50% of total global
emissions, will not sufficiently slow the pace of climate change, thereby
failing to justify the economic costs entailed. Added to this economic
rationale is a worrisome trend toward increased skepticism among U.S.
politicians regarding the most basic and widely established tenets of
climate science, despite the annually compounding evidence of their
veracity.

Whereas the stalemate between these two positions has stalled
the present set of international negotiations, the basic architecture of

the governing Kyoto framework suffers from a more fundamental set of limitations. The first of these is its reliance on traditional modes of international governance to develop the principal response to climate change. A relatively weak mode of societal governance, broadly constructed international agreements generally lack the enforcement authority required to bring about significant changes in the economic activities of participant nations. Evidence of this basic limitation is the general absence of enforcement mechanisms for noncompliance with the Kyoto Protocol. Ratifying nations that fail to meet their emissions-reduction obligations are subject to no direct penalties under the agreement, except for an increase in unenforceable emissions targets for the next agreement period. Given the lack of tangible costs of noncompliance, about one in five Annex I nations are expected to fall short of their obligations by the end of 2012 [4, 5].

A second fundamental limitation of the Kyoto framework is its decoupling of actions regarded as mitigation from those regarded as adaptation. In accordance with the Framework Convention definition of climate change, mitigation is defined as "an anthropogenic intervention to reduce the sources or enhance the sinks of greenhouse gases." By contrast, adaptation is defined as "adjustment in natural or human systems in response to actual or expected climatic stimuli or their effects, which moderates harm or exploits beneficial opportunities" [6]. In short, mitigation is understood as counteracting the physical drivers of climate change, whereas adaptation is understood as enabling an adjustment to resulting changes that are not successfully mitigated. In light of these definitions, mitigation logically precedes adaptation and, in fact, seeks to negate the need for adaptation altogether.

Because successful mitigation eliminates the need for adaptation, the Kyoto framework provides little guidance on how national governments should prepare for the inevitable impacts of climate change. Indeed, the framework fails even to prioritize actions that might serve to enhance population resilience while simultaneously achieving reductions in greenhouse gas emissions. Regional reforestation programs, for example, can assist in meeting emissions targets based on the extent to which such projects sequester carbon dioxide from the atmosphere, but any concurrent benefits in the form of reduced temperatures, soil erosion control, and improved regional flood management are neither recognized nor incentivized under the treaty.

As a result of this decoupling of mitigation and adaptation, the global policy regime overlooks the fact that some approaches to mitigation are more adaptive than others. And born of this oversight is a failure to leverage community self-interest most effectively in the development of emissions-control programs. As the impacts of climate change grow more acute with each passing year, it is adaptation, not carbon control, that is likely to emerge as the principal focus of climate-management policy. In the face of more intense heat waves, crop failures, and destructive storms, management strategies intended to yield a more stable climate many decades into the future are likely to give way to strategies yielding some protective capacity in our own lifetime. As a result, mitigation, as defined by the Framework Convention, will need to be increasingly linked to adaptation to remain viable – an eventuality unanticipated by the present global-policy regime.

Of the Kyoto Protocol's many limitations, it is its failure to harness inherently adaptive land-based approaches to climate management, characterized in Chapter 3 as "land-based mitigation," that is most detrimental to cities. Because only gaseous emissions are recognized as causing human-driven climate change, radiative climate-change agents – the other half of the greenhouse equation – are not the subject of mitigation policy. Governed by the Framework Convention definition of climate change, the Kyoto Protocol gives rise to a management regime through which a reduction in radiation-trapping gases is viewed as climate mitigation but a reduction in the radiation itself is not. In conflict with more than a century of physical climatology, and in direct opposition to the scientific body formed to advise the United Nations, the limited approach to climate mitigation enshrined in the Kyoto Protocol may be the chief impediment to the development of a new global framework. To understand why this is the case, we must consider not only the challenges confronted by the planet's largest cities but also the opportunities presented by its least populated expanses.

The Sahara Desert is among the most extreme environments on Earth. With warm-season temperatures reaching 120°F and few sources of water, the Sahara is across most of its vast expanse uninhabitable by humans. In fact, the Sahara of North Africa is by far the largest mid-latitude land mass that remains almost entirely unsettled and uncultivated by humans. It is precisely its mid-latitude setting that makes the Sahara the focus of one of the most ambitious proposals for managing global climate change to appear in scholarly journals of climate science. If a moisture source can be found, there is the potential to transform the Sahara into a vast biophysical engine for the sequestration of carbon dioxide by mid-century. What is most compelling about the proposal to establish a subtropical forest across the Sahara is not the effort required to water an expansive desert but rather the extent to which human emissions of carbon dioxide could be removed by such a forest. In theory, a Saharan subtropical forest could absorb enough CO_2 each year to *cease altogether* greenhouse-driven climate change, and it could do so without the decommissioning of a single power plant.

If the idea of cultivating a forest in a desert sounds more like science fiction than a plausible mitigation strategy, it is useful to recall that the Sahara has been forested in the past. Indeed, the petrified remains of ancient forests are strewn throughout the desert, attracting tourists to sites in many parts of northern Africa. As recently as 5,000 to 6,000 years ago, the Saharan landscape was a relatively wet environment, populated by lakes, wetlands, dense vegetation, and numerous species of grazing animals on which human populations subsisted [7]. The fossil record suggests that regime shifts between a wet and a dry Sahara are part of a recurring cycle in northern Africa, governed by larger, global climatic cycles and, most recently, influenced by the activities of humans. In fact, there is little in the fossil record to suggest that the Sahara will not at some point transition again to a wet regime or that such a transition may not be hastened through the introduction of water.

As we have seen, the removal of forest-land covers across the Amazon and Australian Outback has played a direct role in the decrease in rainfall across these regions over time. With the widespread loss of tree cover comes a reduction in local evapotranspiration and cloud formation, effectively robbing the atmosphere of the moisture needed for precipitation. Conversely, the introduction of trees to arid landscapes can eventually induce rainfall through a priming of the regional hydrological cycle. The principal challenge confronting a reforestation of today's subtropical deserts is the need for an initial source of water to establish a young forest – the same role played by shifting monsoon patterns in earlier eras. To assess the extent to which an introduction of irrigation water could bring about a more self-sustaining forest, a team of climate scientists is exploring the potential to irrigate massive swaths of the Sahara and Australian Outback through the construction of desalinization plants and development of inland water-delivery systems [8]. Both the economic and environmental implications of such a seemingly infeasible proposal yield important insights for the present climate-policy debate.

From a technological perspective, the construction of desalinization plants and highly efficient water-delivery systems relies entirely on mature technologies. With sufficient investment, such systems could be constructed to establish large-scale tree plantations. It is estimated that the cultivation of expansive forests of Eucalyptus, a rapidly growing species of tree well adapted to arid climates, across uninhabited regions of northern Africa and Australia could sequester as much as 6 to 12 giga-tonnes (Gt) of carbon from the atmosphere annually, with the midpoint of this range exceeding total global emissions from fossil fuels in the present period. Of course, the construction and operation of the irrigation infrastructure itself would entail large investments of energy and associated carbon emissions. Assuming the power required for seawater filtration and delivery is provided by natural gas, a less carbon-intense source of fossil energy than coal, the total carbon-sequestration benefits of desert forestation is diminished by about 20% but still result in sufficient removal of atmospheric CO_2 to substantially, if not completely, offset present annual emissions.

The economic cost of establishing and maintaining sequestration forests across the planet's largest subtropical deserts depends directly on the degree to which such forests can induce their own rainfall. Assuming that 100% of the needed water must be supplied via engineered irrigation – as would be the case during the initial period of forest establishment – the annualized cost of such a system is estimated to be roughly equivalent to a $1 tax per gallon of gasoline. However, because most of the costs relate to the filtration and delivery of irrigation water, the less irrigation needed over time, the less costly the endeavor. To assess the impacts of an irrigated Sahara and Australian Outback on both regional and global precipitation patterns, NASA's ModelE global-climate model was run with and without expansive Eucalyptus forests established throughout these regions. The results of these model runs for global precipitation are presented in Figure 5.2.

As illustrated in Figure 5.2, the regional implications of afforestation in the Sahara and Australian Outback are potentially substantial. Compared to a world without forests in these regions, rainfall increases by about 2 to 4 mm per day across the desert, providing in most areas the 2 to 3 mm per day needed to sustain such a forest. Globally, the establishment of expansive forests in these regions could modestly enhance rainfall in other parts of the planet, such as over the eastern United States, southern Europe, and central to southern Asia. Accompanying these increases in rainfall outside of the afforested areas could be a modest reduction in rainfall, which occurs, based on these model runs at least, almost entirely over the ocean. Overall, the global implications of regional afforestation for rainfall are estimated to be minimal.

Thus, when simulated with the most advanced global-climate models available, afforestation of large subtropical deserts is shown to eventually supply a large percentage of the rainfall needed to sustain the

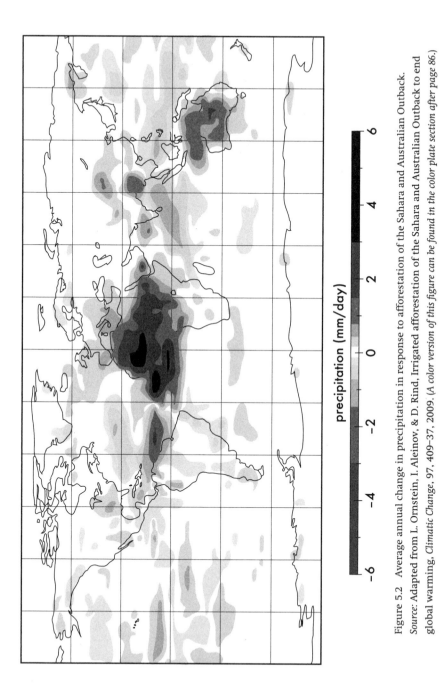

Figure 5.2   Average annual change in precipitation in response to afforestation of the Sahara and Australian Outback.

Source: Adapted from L. Ornstein, I. Aleinov, & D. Rind, Irrigated afforestation of the Sahara and Australian Outback to end global warming, *Climatic Change*, 97, 409–37, 2009. (*A color version of this figure can be found in the color plate section after page 86.*)

forested zones, reducing the required economic investment over time. And, if it is possible to cultivate the expansive reaches of these subtropical deserts, the quantity of carbon dioxide sequestered could negate the need for any other carbon-management strategy, worldwide. If such outcomes are possible, it is easy to grasp why global policymakers would be interested in pursuing this strategy, for it is the only mathematically demonstrated proposal that not only fully offsets annual global emissions of carbon dioxide but also offers the potential to actually draw down atmospheric levels of carbon dioxide over time. It is a proposal that delivers far deeper reductions in atmospheric carbon than any strategy focused on a transition to renewable sources of energy or requiring the development of carbon capture and storage systems for fossil fuels like coal. In addition, it is the only proposal that holds any real potential to deliver these outcomes in the lifetime of present generations.

But a more pressing question, and one too rarely posed by the governing global framework for climate management, concerns if and why the sovereign nations of northern Africa would be interested in assuming responsibility for managing the problem of global-scale climate change. For whereas the benefits of carbon mitigation to the planet as a whole are clear, the benefits of carbon management for individual nations, particularly if those benefits would not be realized for decades or centuries, are less so.

The answer to this question is likely to have very little to do with carbon mitigation at all. Host nations for global forestation would be more likely to pursue such strategies to achieve regional-scale, adaptive benefits related to heat management, drought management, and food production than for carbon control. What is most appealing about land-based mitigation in the form of regional forestation strategies is that in contrast to the dominant approaches to mitigation under the Kyoto Protocol, global and local interests come into close alignment.

What would be the regional benefits of desert afforestation for the host nations of northern Africa and Australia? Figure 5.2 illustrates the first benefit: induced rainfall outside of the immediate zone of afforestation. Huge swaths of marginally productive farmland along the North African coast and within the Sahel region, a climate transition zone between the Saharan desert and equatorial rainforests to the south, would be likely to receive additional rainfall, boosting the production of food on the continent where it is most needed. Along these lines, harvesting rotations within the forest itself, needed to maximize the sequestration of carbon over time, would provide tremendous volumes of wood to be sold as lumber or used as a renewable fuel source for power generation.

In addition to increasing rainfall in an arid environment, extensive afforestation across the Sahara and Australian Outback would have a profound influence on regional temperatures. As illustrated in Figure 5.3, greatly enhanced evapotranspiration would have the effect of lowering regional temperatures by more than 4°C (7°F) throughout much of the forested zones, with associated cooling benefits immediately outside those zones. The global-climate model runs also show only very minimal impacts on temperatures outside the Sahara and Australian Outback, with the light blue and yellow shades in Figure 5.3 indicating temperature changes close to zero. In short, global-climate models reveal no significant, global-scale climate penalties for pursuing large-scale afforestation at subtropical latitudes.

Despite the theoretical potential for substantial benefits to result from a large-scale afforestation of the Sahara and Australian Outback, it is not my intent to endorse this strategy as the solution to the problem of global-climate change. To the contrary, the concentration of global-forestation efforts into one or two massive zones presents a number of significant limitations. The most apparent of these is substitution of one all-encompassing strategy (emissions control only) with another (sequestration only) when both strategies need to be pursued to the greatest extent achievable. Other significant drawbacks include the need for extensive infrastructure to cover vast expanses of the world's largest deserts, the requirement for far more irrigation than would be needed for reforestation projects in less arid regions of the planet, and the imposition of unrecognized risks and externalities from a global management scheme on just a few nations. Indeed, what is arguably more problematic than the concentration of costs resulting from such an undertaking is the geographic concentration of adaptive benefits, which will be greatly needed in closer proximity to a larger proportion of the global population.

What then seems most valuable about this line of research is not its proposal of a silver-bullet solution for global warming but its demonstration of the potential for forest cultivation at low to mid-latitudes to address both the global and regional dimensions of the climate change problem. In response to the initial question posed at the start of this chapter – are land-use strategies an effective means of global carbon management – these results clearly illustrate the extent to which land-use strategies can provide a powerful tool for drawing down global atmospheric carbon. To be most effective as a climate-management strategy, however, and to

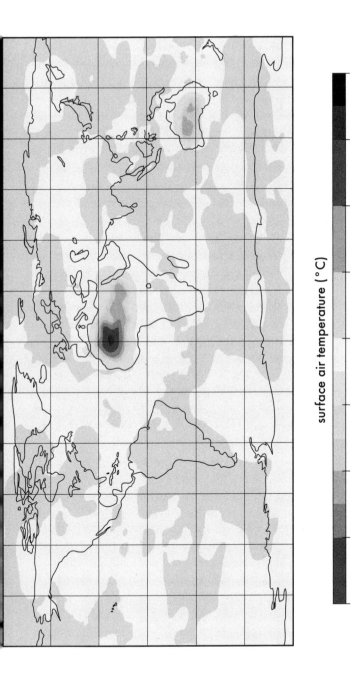

surface air temperature (°C)

Figure 5.3 Average annual change in temperature in response to afforestation of Sahara and Australian Outback.
Source: Adapted from L. Ornstein, I. Aleinov, & D. Rind, Irrigated afforestation of the Sahara and Australian Outback to end global warming, *Climatic Change*, 97, 409–37, 2009. (*A color version of this figure can be found in the color plate section after page 86.*)

be most relevant to the challenges confronted by urban environments, global forestation efforts will need to be targeted not toward one or two massive zones but rather toward the many regions where the adaptive benefits of forest cultivation are needed. Therefore, the principal challenge becomes the development of a policy framework to bring about extensive reforestation in many places around the planet, both rural and urban. It is with respect to this policy goal that the adaptive benefits of reforestation strategies are revealed not as distinct from mitigation but rather as inherently intertwined.

In the Netherlands, perhaps the only issue more politically charged than the dismantling of levees is the construction of new coal-fired power plants. Confronted with rapidly rising costs of oil and natural gas, coupled with a growing intolerance for nuclear power among its citizens, the Dutch government is finding it increasingly challenging to address a rising demand for energy while meeting its emissions reductions obligations under the Kyoto Protocol. In recognition of the infeasibility of meeting those demands through renewable sources alone, construction of a handful of new coal-fired power plants has been proposed in the last decade. To no one's surprise, these proposals have not been embraced by a population well attuned to the risks of climate change. So unpopular is the expansion of a coal-fired plant in Rotterdam, for example, that protesters have in recent years shackled themselves to its coal conveyance systems, effectively disabling the plant. When this tactic failed to derail the project, Greenpeace dispatched anti-whaling ships to blockade the city's harbor to coal shipments, completing the ascendancy of climate change in global environmental politics.

In response to the public outcry, the Dutch government has pursued two general strategies to offset carbon emissions from coal-fired power plants. The first approach is to substitute biomass for coal in power-plant furnaces. Starting in the early 1990s, wood pellets derived from sawdust and lumber recycled from construction projects have been used as a fuel source, accounting for as much as a quarter of the input fuel at some plants today. At other plants, biomass in the form of sludge from wastewater-treatment facilities is burned to generate electricity, further offsetting the need for coal. In both instances, the use of biomass offsets total carbon emissions by substituting a renewable, plant-based source of energy for coal. Because the carbon dioxide released through the burning of wood or sludge is removed from the atmosphere through the subsequent growth of timber plantations or food crops, such fuel sources are largely carbon neutral.

The second strategy has involved reforestation projects, both at home and abroad, to sequester the excess carbon emissions produced by new or expanded coal-fired power plants. Adopting the same principles informing the Sahara afforestation proposal, but on a much smaller scale, the Dutch government is underwriting the reforestation of more than 180,000 hectares of denuded forestland in Malaysia, Uganda, Ecuador, and the Czech Republic with the aim of fully sequestering the emissions of a 600-megawatt power plant (sufficient to power about 1.5 million Dutch homes) in a 25-year period [9]. The Netherlands is undertaking this reforestation project not to meet the requirements of the Kyoto Protocol – the project was put in place before enactment of the protocol – but rather to achieve its own more aggressive emissions-reductions targets established in advance of the protocol's ratification. In the process, the reforestation program is providing an infusion of investment in developing economies, restoring lost habitat to endangered species, and enhancing a range of regional environmental services, such as watershed protection and regional heat mitigation. Most important, the Dutch reforestation program has provided a model for a much more expansive program under the Kyoto framework itself.

The global potential for carbon sequestration through reforestation projects at tropical latitudes – latitudes at which reforestation has been shown to yield the greatest climate-mitigation benefits – is substantial. Worldwide, it has been roughly estimated that 1.3 Gt of carbon could be sequestered each year by the establishment of forests on degraded lands in tropical biomes [10], mitigating about 18% of the total carbon emissions resulting from fossil energy use each year. Because such degraded lands include formerly cropped or forested areas that are often subject to extensive soil erosion, tree planting there takes advantage of lands that are otherwise of limited use in food production, tend to be sparsely populated, and are likely to be providing limited revenue to owners or governments responsible for their management. And, in contrast to the expansive deserts of the Sahara and Australian Outback, degraded lands in tropical biomes enjoy sufficient annual rainfall to regenerate forests without the need for energy-intensive irrigation systems.

From a political perspective, global reforestation programs would seem to offer an ideal alignment of interests between the global North and South. As illustrated in Figure 5.4, virtually all of the tropical-forest regions are found within developing countries, affording these countries a new source of revenue in the form of carbon-sequestration payments that could augment the limited revenues presently generated

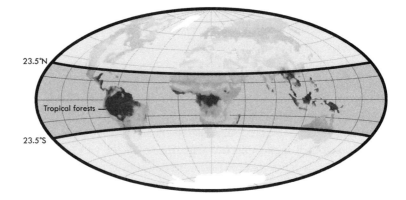

Figure 5.4   Global tropical forests. *Source:* Adapted from R. Simmon, NASA
Earth Observatory.

by unproductive or degraded lands. From the perspective of industrialized countries, forestation projects offer an approach to managing
carbon emissions that is far cheaper than conventional emissions controls. Compared to an estimated cost of about $50 to $80 per ton of
avoided emissions through industrial carbon-capture systems, forest-
management projects can produce the same atmospheric benefit for
a cost of between $5 and $20 per ton of carbon, providing 2.5 to
16 times the mitigation potential of technological controls per dollar spent [11]. With this mitigation comes at least two climate-related
benefits that are unattainable through technological controls: a reduction in local temperatures and the potential not only to stabilize but
also to draw down atmospheric concentrations of carbon dioxide over
time.

   Well aware of the substantial benefits possible through global
reforestation programs, the drafters of the Kyoto Protocol have developed a policy mechanism through which developed nations can obtain
credit toward their carbon-reductions targets through the sponsoring
of reforestation projects in developing nations. The Clean Development Mechanism (CDM) is a program under the Kyoto Protocol through
which Annex I countries may invest in projects in developing countries designed to reduce emissions or enhance the sequestration of carbon beyond levels that would have been achieved absent international
investment. Although most CDM projects to date have taken the form
of energy-related projects in rapidly growing countries such as China
and India, reforestation projects in non-Annex I countries are also eligible for the issuance of carbon credits. However, only a very small

number of approved CDM projects have employed reforestation as a carbon-management strategy, accounting for less than 1% of all CDM projects [12].

The failure of the Kyoto Protocol to bring about substantial investment in global reforestation is among the most significant shortcomings of a framework riddled with policy limitations. The more restrictive policy requirements placed on forestation projects relative to conventional technological controls are a principal impediment to a greater emphasis on reforestation. For example, the magnitude of emissions that may be addressed through international reforestation projects is capped at no more than 1% of the country's 1990 emissions. By contrast, mitigation strategies entailing international investment in industrial mitigation are subject to no such caps, effectively prioritizing these actions over carbon sequestration. Also problematic is the need to demonstrate that no reforestation would have occurred in the absence of a CDM project, a standard virtually impossible to meet.

But the most significant impediment to forestation projects under the Kyoto Protocol has been the ineligibility of such projects for permanent carbon credits. Because of concerns over the potential for forest fires or infestations to release sequestered carbon at a future date, the carbon captured through reforestation projects is eligible only for temporary carbon credits, which, in the eyes of investors, are clearly less valuable than credits that may be traded in perpetuity. Confronted with a reduced market potential to sell carbon credits earned through reforestation projects at a future date, many fewer governments or industries have pursued this option, relying instead on energy-related carbon programs that yield almost no adaptive benefits to host communities.

Related to the inability of the Kyoto Protocol to stimulate widespread reforestation activities, and substantively more problematic, is the failure of global climate policy to address the continuing problem of *de*forestation. Annual global deforestation, which is occurring almost entirely in tropical zones, is undermining the international effort to slow annual emissions to a staggering extent. It is estimated that a cessation of rainforest destruction in just two countries, Brazil and Indonesia, would by itself bring the world four-fifths of the way to meeting the cumulative targets of the Kyoto Protocol [13]. Stated another way, 80% of the global emissions reductions called for under the Kyoto Protocol, a massive policy effort requiring the participation of every developed nation on Earth save one and the investment of billions of dollars in technological emission controls, are erased annually by the continued destruction of rainforests in Brazil and Indonesia. What is perhaps most discouraging about these numbers is the often low-value uses for which these lands are being cleared. In Brazil, for example, most rainforest destruction is not carried out to harvest the wood, which is simply burned on site, extinguishing in the process carbon stores built over eons, but rather to free up land for cattle ranching, which itself produces vast quantities of additional greenhouse gas emissions.

In light of the extent to which rainforest destruction is fueling global climate change, it seems obvious that a major thrust of international climate policy should be focused on reducing the rate of global deforestation. Yet, incredibly, the Kyoto Protocol as presently structured would not award a single carbon credit in exchange for a full cessation of rainforest destruction worldwide. Predicated on a definition of climate change that ignores the land-surface drivers of the greenhouse effect and strongly oriented toward technological control strategies, the Kyoto framework includes no mandatory provisions to slow global deforestation. The result is a programmatic blind spot that renders even the narrowly circumscribed objective of the Kyoto Protocol – a reduction in the atmospheric accumulation of carbon dioxide – almost impossible to achieve.

So, in response to the second question posed by this chapter – *to what extent are land-use strategies being pursued under the international climate-policy framework* – the current policy regime has proved largely ineffective in managing carbon through land-use programs focused on forest protection and regeneration. Were a successor agreement to the Kyoto Protocol not only to enable the investment by Annex I countries in forest protection and regeneration but also to strongly incentivize it, the international community would find the task of stabilizing and ultimately reducing atmospheric concentrations of greenhouse gases far more feasible to achieve. In theory, the combined strategies of protecting what remains of tropical forests and of reforesting degraded tropical lands could reduce or offset global carbon emissions by about 3 Gt of carbon per year, accounting for more than a third of current global emissions.[2] To recommend such a strategy is not to place an undue burden on tropical nations; to the contrary, the industrialized world should be willing to generously outbid any competing use for these lands because forestation strategies will continue to offer one of the most cost-effective means of drawing down atmospheric carbon.

[2] This figure assumes that 1.6 Gt C per year can be avoided through a cessation of tropical deforestation [14] and that 1.3 Gt C of additional carbon sequestration can be brought about through forest regeneration at tropical latitudes [10].

Yet, even greater are the potential benefits to host nations of reforestation projects, for such projects would bring with them a new and significant source of revenue in two forms: revenue in the form of annual sequestration payments from Annex I nations and revenue derived from the sustainable harvesting of forest products, which will be required over time to maximize carbon uptake. Equally valuable to these regions would be the environmental dividends of forest protection and regeneration, which may take the form of enhanced rainfall, reduced flooding, and, of course, reduced temperatures. It is with respect to these adaptive benefits of land-based mitigation strategies that the international policy regime carries the greatest potential to support the climate-management needs of cities. For if the policy framework can be restructured to more effectively bring about forest protection and regeneration in rural areas, these same mechanisms can be employed in urban environments, where tree planting yields dividends both in the form of carbon sequestration and a reduction in the energy consumption of buildings.

In short, the global policy regime can most effectively achieve its goal of carbon management through the promotion of strategies that yield adaptive benefits to the regions in which these strategies are instituted. Referred to herein as "adaptive mitigation," such strategies are defined as climate-management activities designed to reduce the global greenhouse effect, through the control of gaseous and/or land-surface drivers, while producing regional climate-related benefits in the form of heat management, flood management, enhanced agricultural resilience, or other adaptive benefits. The promotion of adaptive mitigation is important not only due to its ability to address the near-term impacts of climate change on regional populations but also for its potential to exceed the magnitude of carbon reductions achievable through nonadaptive mitigation alone. To better understand how adaptive mitigation can ultimately lead to deeper carbon reductions than conventional mitigation, and why such strategies are the key to linking the global and urban climate-management frameworks, we must turn our attention to one of the least urbanized regions in the world: sub-Saharan Africa.

It is noteworthy that a final agreement on the Kyoto Protocol was reached during one of the most extreme global climate events of the 20th century. During the December 1997 meeting of the parties to the UN Framework Convention on Climate Change in Kyoto, Japan, the most intense El Niño event ever measured had reached its height, producing extreme rainfall in some regions of the planet and intense drought in others.

One of the regions plagued by extreme flooding was the central African nation of Uganda, where the combination of mountainous topography and heavy rainfall would result in hundreds of drowning deaths and the displacement of more than 150,000 flood victims – a disaster-relief challenge compounded by the washing away of almost half of the country's major trucking routes. Of greater long-term consequence than the disrupted transport of food aid, however, would be the severe erosion of hillside and valley croplands, the result of heavy and prolonged rainfall, which has deepened poverty and food insecurity in many parts of Uganda.

The long-term outcome in one western Ugandan community, a small village named Kyantobi (which translates literally as "waterlogged"), would be different, however, and rooted in Kyantobi's experience is an environmental narrative that could in some ways reshape the global response to climate change. Confronted with the loss not only of a season's crops but also of much of the productive soils needed for continued food production, a group of villagers solicited assistance from the Agroforestry Research Networks for Africa (AFRENA), an African nongovernmental organization focused on rural food security. As the organization's name suggests, AFRENA promotes an integration of forestry and agriculture to enhance agricultural yields in extreme environments, a traditional agricultural practice abandoned over time in favor of the use of industrial fertilizers and engineered seeds. Because climate change is rendering environmental conditions more extreme in many regions of the planet, traditional modes of food cultivation are being found to enhance the resilience of croplands to highly variable conditions of saturation and drought – breathing new life in the process into an ancient system of agriculture that is far less energy intensive.

The practice of interspersing trees among food crops is believed to be as old as the practice of agriculture itself. The integration of trees into cropland after several years of cultivation was a common practice in medieval Europe, as was the practice of integrating orchard trees into sheep pasture. In the tropics, trees, shrubs, and crops have long been intermixed to mimic the vertical plant structure of the rainforest because such a structure was known to shield low-growing crops from heavy, soil-eroding rainfall. Also desirable was the fertilizing effect of trees on food crops. The practice of alley cropping, a type of agroforestry through which corn, beans, and other staple crops are planted between rows of trees, can increase crop yields over mono-cropping practices by adding to the soil a highly mineralized form of nitrogen. Studies of maize farming in Costa Rica, for example, have found crop yields using alley cropping to be 120% higher than those of single crops, despite a reduction in the available land for maize planting caused by tree shading [10].

Yet, undoubtedly the greatest advantage of tree planting in regions with highly erodible soils is the stabilization of soil through root development. Because trees and most food crops generally occupy different growth zones above and below ground, tree growth can both fertilize and stabilize soils without competing with crops for nutrients or other resources. In recognition of this clear benefit over conventional farming practices, AFRENA helped reintroduce basic agroforestry techniques into the village of Kyantobi in the aftermath of the 1997 floods, largely through the provision of funds to purchase seedlings to be planted along steep hillsides. The result has been the gradual development of a diverse cropping system that is far more resilient to the ongoing cycle of intense rain and drought than farming practices requiring the removal of trees. Through the use of fruit trees and species of shrubs that serve as fodder for livestock, villagers have supplemented their production of staple crops with additional sources of nutrition and income. The availability of wood for fuel use and construction provides another source of income during seasons in which crops fail.

In addition to the increased diversity of food and woody resources now produced in Kyantobi, the most directly measured benefit of agroforestry is revealed in changes to average household income. By 2004, just six years after the reintroduction of agroforestry practices to the village, average household income had increased from about $9 to $29 a month – a more than threefold increase that has carried the average household across the Ugandan poverty threshold [15]. So successful has been the integration of trees into farming practices that many villagers have established their own tree nurseries to meet growing regional demands for tree planting, representing another source of income. But perhaps the most unexpected source of new income from agroforestry is also the most instructive to global climate policy: requests for demonstrations of Kyantobi's cropping techniques have become so frequent that villagers today are able to command a hefty fee for guided tours.

Such a fee is a small price to pay for a better understanding of the linkages between global climate policy and local food security. For, while the agroforestry practices reintroduced into Kyantobi and a growing number of subsistence farming communities are not undertaken with the principal intent of enhancing carbon uptake, they do so nonetheless and with tremendous global potential. As illustrated in Figure 5.5, studies have found the integration of trees and crops to increase the quantity of carbon sequestered per hectare of cultivated land from less than

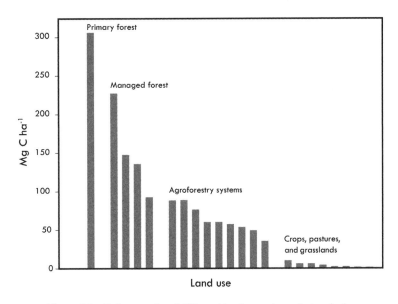

Figure 5.5   Carbon stocks of different land-cover types in tropical zones.
Source: Adapted from Verchot, L., Noordwijk, M., Kandji, S., et al., Climate change: Linking adaptation and mitigation through agroforestry, *Mitigation and Adaptation Strategies for Global Change*, 12, 901–18, 2007, Fig. 2.

10 Mg for sole cropping systems to between 50 and 75 Mg [16]. Whereas this density of carbon sequestration is considerably less than that of forested land, which can reach 300 Mg per hectare, the estimated area of heavily eroded or otherwise degraded agricultural lands at low to mid-latitudes is immense, extending well beyond the wet, equatorial zone of western Uganda. Hundreds of miles to the north of Kyantobi, for example, in the Sahel region of Africa, agroforestry is being introduced as a technique for cultivating crops in a hot and arid environment. In this environment, trees are valued less for their ability to protect soils from flooding than for their ability to retain soil moisture between infrequent rain events and to moderate temperature extremes through shading and evapotranspiration.

Worldwide, a tremendous area of marginal land used for subsistence cropping at low to mid-latitudes, in both wet and arid environments, could be converted from sole cropping to agroforestry. Doing so could increase the global potential for carbon sequestration by another 2.1 Gt per year [10], equivalent to about 25% of total global carbon emissions presently released from the burning of fossil fuels.

Together, the global potential for carbon sequestration from a cessation of tropical deforestation, reforestation of lands not currently engaged in cropping, and the introduction of agroforestry into marginal agricultural lands could account for about 60% of total annual fossil-fuel emissions – bringing us much of the way to what could theoretically be achieved through reforestation of the Sahara and Australian Outback but without the needed infrastructure investment or additional carbon emissions. Most important, such initiatives could greatly increase the resilience of local populations to multiple threats of climate change, ranging from flooding and extreme heat to declining food security.

Of course, the extent to which global investment in agroforestry and other forestation initiatives could approach theoretically possible levels of carbon uptake is limited by many factors. In some areas, the extent of land degradation may be too great to support a reintroduction of trees, at least not without extraordinary efforts. In other areas, some proportion of reestablished forests will burn or succumb to disease each year, offsetting accumulated stocks of carbon. In some regions, international investment in agricultural activities may be unwelcome because of concerns that such investments might carry with them unanticipated obligations or the stigma of neo-colonialism. However, many regions of the tropical world would embrace financial and other forms of international assistance aimed toward forest protection, regeneration, and agroforestry projects as a means of combating poverty and enhancing environmental resilience in areas that otherwise carry few prospects for economic development. Even if these areas account for as little as one half or one third of the theoretical total, forestation projects there carry the potential to achieve greater reductions in atmospheric carbon each year than would be achieved through full implementation of the Kyoto Protocol. In this sense, adaptive mitigation in the form of avoided deforestation, reforestation, and agroforestry carries great potential to augment the magnitude of carbon reductions achievable through conventional mitigation alone.

Yet, here again, the limited orientation of the UNFCCC and its associated protocols overlooks the potential for land-based mitigation strategies to play a significant role in global carbon management. As with a slowing of tropical deforestation, agroforestry programs that could sequester millions of tons of carbon from the atmosphere each year, while at the same time enhancing the resilience of subsistence food systems in extreme environments, are not eligible for emissions-reduction credits under the Kyoto Protocol – an oversight that not only deepens carbon reductions needed elsewhere but also fails to address the growing adaptation needs of highly vulnerable populations.

Therefore, the principal lesson from Kyantobi's experience is not that agroforestry and related efforts can by themselves solve the problem of global climate change. It is rather that the scale of carbon reductions needed to stabilize global climate – reductions far deeper than called for by the Kyoto Protocol – will only become feasible if married to the political, economic, and social needs of the communities bringing about these reductions. At present, the global policy framework for managing climate change places the responsibility for achieving carbon reductions on too few nations while offering too little in return for their efforts: a slowing of the rate of planetary warming many decades into the future, with unknown benefits, if any, accruing during the lifetime of those incurring the costs of carbon management. Such a temporal and spatial discontinuity between cost and benefit grows ever more untenable in environments already confronted with extreme climate-related challenges – environments found not only in rural Africa but also in large cities confronted with a steadily rising frequency and intensity of extreme heat. It is in the latter of these environments – large cities of both the developing and developed worlds – that an emissions-only approach to climate management is proving most counterproductive.

On a hot summer afternoon in virtually any major American city, the temperature of an unshaded asphalt street can reach a sweltering 130°F. Painful to the unshod foot and deadly to the wayward worm, solar-heated streets represent more than a source of discomfort for those who cross them; in most cities, solar-heated streets are a principal driver of climate change. The reason for this is that the streets and surface parking lots found in downtown districts often occupy more physical area than the buildings they are designed to serve. Covering many thousands of acres of paved area across a metropolitan region, streets and parking lots act as a massive solar collector in cities, absorbing prodigious amounts of thermal energy that can only be returned to the atmosphere in the form of sensible heat or greenhouse-enhancing radiation. Although there is potential to cool paved surfaces through the planting of street trees and the use of moderately reflective paving aggregates, a more ingenious solution is possible: harness the absorbed solar energy to generate electricity.

The idea has much to recommend it. Streets and other paved surfaces are a ubiquitous source of renewable energy found throughout urbanized areas. In contrast to traditional solar-collection systems, streets retain their heat energy well after sunset, enabling energy to be produced for a longer period each day. Most streets are publicly owned, so the acquisition of high-value urban land is not required for the siting of a new energy system. In addition, the need to repave streets every decade or so provides an ongoing cycle of maintenance during which subsurface heat-collection systems could be installed. Were thermoconducting coils to be embedded within newly laid asphalt, a circulating fluid could be heated and used to drive an electrical generator or used directly as a heat source, a relatively low-tech process already in operation in small-scale demonstration projects.

What these demonstration projects reveal is that roadway energy systems provide not only a cheap source of energy but also an effective means of managing excess heat in cities. In one study based in New England, for example, an embedded thermal-collection system was found to lower asphalt temperatures by an average of 12°F, translating into cooler air temperatures in proximity to the paved surface [17]. At a second demonstration site in the Netherlands, a roadway energy system is being used to provide heat to a 70-unit apartment building, where it cuts fossil fuel use and associated greenhouse gas emissions by more than 50% [18]. Roadway energy projects in large cities of the developed world, therefore, have something in common with agroforestry projects in rural Africa: both are clear examples of adaptive mitigation, serving to enhance local climate resilience while reducing atmospheric carbon. As such, these strategies share a second commonality: neither approach is adequately incentivized by the global policy framework for climate-change management.

If rural areas carry the greatest global potential to bring about significant reductions in atmospheric carbon through the pursuit of adaptive mitigation strategies, it is urban areas that have the most to lose from failing to do so. Home to the largest and most rapidly growing share of the global population, cities exhibit a profound vulnerability to climate change not only because of their dense populations but also due to the growing extremity of urban environments. It is in these environments that climate-management regimes should be found in their most robust form, and yet this is almost never the case. Today, no major city worldwide has developed a comprehensive administrative structure to combat the principal drivers of climate change within its own boundaries. This fact, more than any other, illustrates the great significance of the international regime for climate management to the fate of cities because it is the architecture of the UN Framework Convention and its associated protocols that arguably has most impeded the emergence of adaptive, land-based mitigation efforts in cities.

The international climate-policy framework has perhaps most affected climate management in cities by failing to directly engage them. In keeping with the structure of the United Nations, only national governments have a direct administrative role in meeting the carbon-reduction targets of the Kyoto Protocol. In response to the protocol's strong orientation toward reducing atmospheric carbon through technological emissions reductions alone, national governments have most commonly sought to regulate the emissions activities of large industrial emitters, as opposed to the land-use activities of cities or individual landowners.

To address this policy vacuum, nongovernmental organizations, such as the International Council for Local Environmental Initiatives (ICLEI) – now known as Local Governments for Sustainability – have emerged to advise urban governments on voluntary actions that may be taken to address climate change. Although ICLEI works with cities in an advisory capacity on a number of issues related to urban sustainability, it has also become the leading international body advocating for climate management in cities, with the bulk of its efforts since its founding in the early 1990s focused on emissions-control programs at the urban scale. In recognition of the extent to which urban populations are already vulnerable to climate change, ICLEI has recently begun advocating for more aggressive climate adaptation. In doing so, ICLEI is positioning itself as one of the only global organizations actively working to promote climate adaptation alongside mitigation. Yet, the organization's primary focus centers on local actions to mitigate and adapt to global-scale drivers of climate change rather than to the local- and regional-scale mechanisms that are the primary drivers of extreme heat in urban environments.

In the United States, the leading effort to coordinate urban-scale climate management has taken the form of the U.S. Mayors Climate Protection Agreement, a voluntary agreement spearheaded by Greg Nickels, Seattle's mayor from 2002 to 2009, and currently signed by the mayors of more than 1,000 U.S. cities. Through this agreement, urban executives pledge to meet the emissions reduction targets of the Kyoto Protocol, requiring a reduction in municipal emissions to a level 7% below 1990 emissions.[3] In response to the Climate Protection Agreement, the efforts of ICLEI, and passage of some state-level legislation requiring climate-change planning, many large cities have developed formal climate action plans designed to address climate change in the absence of a national management program. In adopting Kyoto targets as a voluntary

---

[3] Had the United States ratified the Kyoto Protocol, the country's emission-reduction target would have been 7% below 1990 emissions.

goal, however, municipal governments have also generally adopted an emissions-only approach to local climate management, failing in the process to address the principal drivers of urban-scale climate change at work in their own jurisdictions.

To assess the extent to which climate action plans in place in large U.S. cities are designed to address local-scale, non-emissions-based drivers of climate change, the Urban Climate Lab at the Georgia Institute of Technology reviewed all such plans available for the same 50 cities included in our urban climate trends analysis (Figure 3.7). The results of this analysis are illustrated in Figure 5.6.

One notable finding is that a significant number of large U.S. cities are not planning for climate change in any formal capacity. Implementing no strategies presented in Figure 5.6, a quarter of those cities surveyed have not adopted climate action plans designed either to reduce municipal emissions of greenhouse gases or to mitigate the local-scale drivers of rising temperatures. In these cities, no planning is being undertaken at the local, state, or national levels to slow or prepare for the effects of climate change.[4] A significant number of these cities are found in the southwestern United States, where the impacts of climate change on water resources and extreme heat are already widely apparent.

For those cities having developed climate action plans, we scored those plans based on their inclusion of one or more of three categories of strategies: (1) tree planting or other vegetation-enhancement strategies; (2) albedo-enhancement strategies incorporating cool-roofing or cool-paving techniques; and (3) energy-efficiency programs designed to reduce greenhouse gas emissions or waste-heat production from municipal buildings, vehicles, or operations. These strategies are presented in separate columns in Figure 5.6. We next scored each strategy based on one of two purposes stated in the plans: local-scale heat management or emissions reductions designed to address the global greenhouse effect. Strategies included in any plan with the stated intent of reducing urban temperatures, whether characterized as a primary or secondary benefit of the strategy, are identified in Figure 5.6 with a closed circle. Strategies included in any plan with the stated intent of reducing greenhouse gases only are identified with an open circle.

Categorized in this fashion, the results of our survey suggest that only a handful of large U.S. cities are actively taking steps designed to

---

[4] Both municipal plans and state plans with provisions for cities were included in the study. There are no federal requirements for climate management in U.S. cities.

City	Vegetation Enhancement	Albedo Enhancement	Energy Efficiency
Akron			
Albany	o	o	o
Albuquerque	●	●	o
Allentown	o	o	o
Atlanta			o
Austin	o		o
Baltimore	o		o
Baton Rouge			
Boise	●	●	o
Buffalo			o
Charlotte	o		o
Chicago	●	●	o
Columbia			
Columbus			
Des Moines			o
Detroit	o		o
El Paso			
Greensboro	o		o
Hartford	●		o
Houston			o
Indianapolis			
Jacksonville	o		o
Knoxville			o
Las Vegas			o
Louisville	●	●	o
Milwaukee	o		o
Minneapolis	●		o
Nashville			o
New Orleans	o		o
New York	●		o
Norfolk	o		o
Oklahoma City			
Orlando	o		o
Philadelphia	●	o	o
Phoenix			o
Pittsburgh	●	o	o
Portland	●		o
Providence			
Richmond	o		o
Rochester			o
Salt Lake City	o		o
San Antonio			
Seattle	o		o
St. Louis			o
Syracuse			o
Tampa	o		o
Toledo			
Tulsa			
Washington, DC			
Wichita			

Figure 5.6   Climate-management strategies of large U.S. cities.
Note: Closed circles denote strategies adopted for the purpose of heat management; open circles denote strategies adopted for the purpose of emissions reduction. Source: Stone, B., Vargo, J., & Habeeb, D. Managing climate change in cities: Will climate action plans work? Landscape and Urban Planning, in review.

address local-scale drivers of climate change. Of the 50 cities included in the survey, only 10 (20%) are pursuing strategies designed to increase vegetative cover or surface reflectivity as a means of reducing ambient temperatures. Although all cities that have adopted climate action plans are pursuing energy-efficiency strategies, no plan cites waste-heat reduction as a goal of these efficiency improvements. In concert, these findings suggest that very few U.S. cities are focused on reducing the occurrence of extreme heat over the next several decades. Consistent with the orientation of global climate policy, most municipal governments are pursuing strategies designed to reduce the emissions of greenhouse gases alone, which by themselves will yield no protective benefits to cities and, even if pursued globally, will not yield protective benefits for many decades, if not centuries. In short, most large U.S. cities are taking no steps designed to counteract the principal threat of climate change to urban populations in the present period: extreme heat.

The good news is that many cities are pursuing emission-reduction strategies that carry the unintended benefit of mitigating local-scale drivers of warming as well. In recognition of the benefits of tree planting and reflective roofing for cooling buildings, thereby reducing energy consumption for air conditioning, more than half of the cities surveyed have included provisions in climate action plans for these strategies – strategies that also serve to reduce ambient temperatures. Although the limited spatial extent of these strategies – typically focused on municipal buildings only – is unlikely to measurably slow the pace of warming in large cities, such approaches at least constitute the beginnings of a more locally adaptive approach to climate mitigation.

The nature of climate-management activities underway in large U.S. cities suggests that despite the failure of the U.S. Congress to ratify the Kyoto Protocol, the basic policy orientation of the international framework is playing a major role in shaping climate policy at the local level. Yet, implicit in the adoption of a Kyoto-like approach to climate management at the local level is a globalized orientation to climate change that is not particularly well suited to the unique climatic conditions of cities. Rooted in a definition of climate change that ignores land-based drivers of warming, the global orientation to climate management overlooks those drivers at the very scale at which they are dominant. The result is an approach to climate management in cities that is only minimally responsive to the near-term public health risks of locally amplified warming trends.

Equally important, the adoption of an emissions-only orientation to climate management in cities has not proven particularly effective in reducing urban emissions. Despite participation in programs like the Mayors Climate Protection Agreement and the drafting of climate action plans, emissions in U.S. cities have been steadily rising, not falling. The primary reason emissions are rising rather than falling, of course, is the absence of a legally binding commitment to reduce emissions in the United States. But also significant is the decoupling of local and global interests through an emissions-only orientation to climate management. As long as global mitigation activities are decoupled from local climate resilience, there is little incentive to invest in such strategies absent a national mandate.

On this point, it is the globalization of the climate-change problem that in many ways impedes further progress at the urban scale. Fundamental to its decision to characterize the problem of climate change as a problem of greenhouse gas emissions alone was an implicit decision on the part of the United Nations to understand climate change in global terms only. For, whereas the land-surface changes that have compounded radiant emissions and fueled warming at the regional scale fall wholly within the administrative mandate of individual nations, diffuse gaseous emissions, transported aloft by global air currents, fall well beyond the control of any single community, nation, or even continent. Residing outside the management capacity of individual governments, climate change so defined assumes the latent imperative of other globalized problems, such as hunger, poverty, or declining biodiversity – problems for which the vastness of scope becomes itself an impediment to action.

A more effective policy framework would seek to understand climate change in regional rather than global terms, marshaling in the process not only the established management capacity of individual governments but also the unmediated self-interest of affected populations. Such an approach would set region-specific emission goals but would prioritize management actions designed to yield adaptive benefits, such as heat-island reduction or improved flood control, in concert with reduced emissions. In support of these goals, scientific assessments would need to be carried out at the regional level, requiring a shift from the global-scale analyses performed by the IPCC and enabling both the land-surface and atmospheric drivers of regional-scale climate change to be measured.

Importantly, a regional-scale framework for climate management would draw on the unique regulatory powers of states and cities, such as the authority to control the horizontal expanse of metropolitan growth or to protect regional forest canopies; these strategies hold tremendous potential to slow the pace of warming in urbanized regions. Such a framework would make explicit the linkages between regional climate stressors and regional climate impacts, downscaling the problem of climate change to a spatial extent at which local communities are empowered to intervene.

Although such an approach would elevate the significance of local adaptation in urban climate planning, it need not do so at the expense of mitigation. To the contrary, an emphasis on adaptive mitigation would carry great potential to bring about steeper reductions in urban emissions than are being achieved at present. The simple reason for this is that land-based mitigation strategies designed to reduce ambient temperatures in cities also serve to reduce energy demand in buildings and associated carbon emissions. Such a conclusion is supported by a large number of studies on heat-island formation and building energy demand. In the most comprehensive of these, tree planting and the use of reflective roofing on residential parcels in 240 regions across the United States were found to reduce the demand for cooling energy by between 8% and 25% [19], representing a more substantial reduction in local emissions than achievable through programs focused on municipal buildings only. For buildings incorporating green roofs or green walls, energy consumption is reduced in both summer and winter through enhanced building insulation, further reducing carbon emissions.

Also significant is the potential for tree planting in and around cities to enhance regional carbon sequestration. In recognition of this potential, the Kyoto Protocol identifies urban-scale revegetation initiatives as a viable strategy for meeting national emissions-reductions targets under the treaty. Consistent with the protocol's limitations on forest-management strategies in general, however, additional accounting requirements are imposed on revegetation strategies, and only a handful of countries have elected to pursue this approach. In addition, Annex I nations cannot receive carbon credits through investment in the urban-vegetation strategies of developing nations.

Therefore, in cities, as with tropical regions of the developing world, the global policy regime for climate management provides only limited incentives for the achievement of carbon reductions through inherently adaptive, land-based mitigation strategies. Whereas municipal governments are fully empowered to implement such strategies on their own, the Framework Convention and its associated protocols are, in many respects, a significant impediment to local-scale climate management. With this issue in mind, the concluding section of this chapter considers how the global climate-policy regime might be restructured to better address the needs of urban environments. Although my focus is explicitly on cities, the crux of these recommendations is concerned with shifting the primary emphasis of climate science and policy from the scale of the planet as a whole to the scale of regions, regardless of the extent of regional urbanization. As argued throughout this book, it is at the regional scale that both the principal drivers and principal effects of climate change are playing out. Thus, it is at the regional scale that we must focus on developing management strategies responsive to both the emissions-related and adaptive needs of communities already confronted with profound shifts in climate.

The most populous cities are well positioned to moderate the pace and extent of climate change within their own boundaries. Typically hundreds of square miles in area, large cities and their metropolitan environs create a climatic footprint sufficiently large to influence temperature, humidity, wind speed, and even rainfall in their own jurisdictions. Consequently, cities are not simply bystanders to a global environmental phenomenon outside of their control but also are active agents in the process of regional-scale climate change. Municipal governments, neighborhood-scale organizations, and even individual property owners can take actions to modify the thermal and hydrological characteristics of the urban environment that are contributing to rising temperatures in their own communities – a class of actions I have referred to throughout this book as land-based mitigation. As the incidence of extreme heat grows ever more frequent in cities, regional-scale, land-based mitigation will become a principal focus of climate-change management, with or without a restructuring of global climate policy to address these issues.

The third and final question posed at the start of the chapter – *how can the international policy framework be restructured to more effectively employ land use as a tool for climate management, particularly at the scale of cities* – is the focus of this concluding section of the book and represents a significant policy challenge. For it may reasonably be argued that regional and national governing institutions are best positioned to address climate-change agents that are predominantly regional in their origins and impacts; therefore, the international framework should focus on global-scale phenomena alone. But this argument overlooks a key benefit of land-based mitigation: the global community potentially has as much to gain from the institution of such strategies as do urbanized regions. Impeded from achieving more substantial cuts in global emissions because of a long-standing impasse between developed and developing nations, the international policy framework will need greater flexibility if it is to secure binding commitments from the United States, China, India, and Brazil – countries that account for about half of annual global emissions but are subject to no emissions obligations under the

Kyoto Protocol. A broadening of the tools available to meet national obligations under a global-climate treaty could lead to a reassessment of binding commitments by one or more of these countries, particularly if it augments the flow of resources from North to South.

Although a discourse on the programmatic details of a revised global-climate treaty well exceeds the scope of this book, the general thrust of policy changes needed to bring about more effective climate management in cities can be outlined. What follows in the remainder of this chapter is a brief discussion of four recommended changes to the existing global climate-policy framework that could accelerate the implementation of land-based mitigation strategies in both urban settings and rural environments confronted with rapid land-use change. I cannot claim any of these ideas as my own because components of each have been the focus of policy debates in the climate literature or treaty negotiations, although generally not for the express purpose of promoting land-based mitigation. In addition, each recommended change to the existing policy framework leaves unaddressed significant programmatic details that can only be resolved through a formal legislative process. All of these ideas stem from what is now a well-established body of scientific work on the role of land use in climate change, much of which has been explored in the pages of this book.

1.   *Broaden the definition of climate change adopted by the UN Framework Convention on Climate Change to encompass the land-surface drivers of climate change.*

Among the range of policy changes to be considered, this first recommendation is perhaps the most easily accomplished, for it only requires adoption of the climate-change definition proposed by the IPCC, the scientific body established by the UN for the very purpose of understanding the phenomenon. The decision by parties to the UN Framework Convention to recognize only the gaseous agents of climate change has given rise to a management framework that overlooks the most powerful drivers of climate change at the urban scale. This decision has framed the problem of climate change in global terms only because only the more narrow problem of greenhouse gas emissions can be effectively managed at this scale. In concert, these two byproducts of the Framework Convention definition of climate change – an exclusive focus on emissions reduction and an understanding of climate change in global terms only – have contributed to a distancing of the climate problem from our everyday lives and from the administrative purview of local and regional governing institutions. Perhaps more than any other policy-related decision, it is the adoption of a scientifically incomplete definition of climate change that has impeded more aggressive action by local governing institutions in managing the rapid warming already underway in large cities.

A broadening of the Framework Convention to account for the land-surface drivers of climate change would open the door to a wider set of approaches to moderating shifts in climate, particularly at the regional scale. Although land-based mitigation will not be undertaken in cities for the exclusive purpose of reducing atmospheric carbon, heat-management strategies can yield significant reductions in building energy consumption and associated emissions while further enabling regional carbon sequestration through vegetation enhancement. By the same token, broader investment in tropical reforestation for the purpose of regional climate management and food production could also yield substantial co-benefits in the form of enhanced carbon sequestration. Revising the Framework Convention to encompass all physical drivers of climate change would expand management activities in ways highly beneficial to rapidly warming regions while advancing the global policy goal of emissions control.

2.   *Adopt the regional scale as the primary focus of scientific assessments of climate change.*

Recognition of the land-surface drivers of climate change necessitates a greater emphasis on regional-scale climate phenomena in climate science and policy. As discussed in Chapter 2, although shifts in albedo and the surface-energy balance carry profound implications for climate at the urban and regional scales, the impacts of such shifts are often obscured by global-scale metrics of climate change, such as the oft-cited mean annual change in global temperature. Because land mass occupies only a third of the planetary surface, land-based drivers of climate change are mathematically diminished when averaged over the planet as a whole. Yet, this statistical artifact in no way diminishes the impact of land-use change in the regions in which it is occurring. If management programs are to be developed to address regional-scale climate phenomena, region-specific scientific assessments will be needed to gauge the principal drivers of warming trends at this scale, as well as the effectiveness of land-based mitigation strategies in slowing these trends.

Greater recognition of the land-surface drivers of climate change would require the IPCC to develop more detailed assessments of climate change at the regional level. Although such analyses have been carried out for large-scale regions in past assessment reports – typically at the continental scale – these analyses are too general in focus and large in geographic area to yield meaningful information for the climate-management activities of state or municipal governments. Often lacking the technical expertise to carry out such assessments on their own, state and municipal governments would benefit greatly from detailed scientific assessments developed for the spatially resolved and climatologically distinct zones in which they are located. Such analyses would create a stronger linkage between science and policy than presently exists in the globally oriented scientific-assessment process and would further improve the precision of global-scale analyses through the aggregation of more highly resolved assessments at the regional level.

3.   *Establish forest protection and reforestation commitments for participating nations.*

Accounting for a tonnage of annual carbon emissions equivalent to the global transportation sector, ongoing deforestation activities in tropical zones constitute an immense source of greenhouse gas emissions that could be addressed through land-based mitigation strategies. In response to this need, global climate-change agreements have been expanded in recent years to call for voluntary investment in forest protection, a strategy known formally as Reducing Emissions from Deforestation and Forest Degradation (REDD). Currently underway in a few nations on an experimental basis, REDD programs represent an important advancement in global climate policy and hold the potential to play a more significant role in the successor agreement to the Kyoto Protocol. To be most effective, however, forest-management programs will need to be funded separately from the carbon-trading schemes instituted under the Kyoto framework, and they will need to have an urban component.

The funding of forest protection and regeneration through carbon markets is problematic in several respects. As previously discussed, the potential for carbon removed through forest projects to be released back into the atmosphere over time leads to a devaluing of these forest-related carbon credits relative to those issued through other carbon-reduction strategies. The creation of a new class of carbon-reduction credits that cannot be traded in perpetuity has rendered forest-related projects less economically attractive than other mitigation strategies – a fact clearly attested to by the small number of reforestation projects undertaken to date.

Also problematic is the lack of precision with which emissions reductions can be estimated from forest-management strategies. In contrast to the mechanical workings of industrial systems, ecological processes are subject to constant fluctuations in response to environmental conditions, undermining the precision with which carbon sequestration can be estimated. Likewise, there is little reliable basis to determine how much deforestation (or forest regeneration) would have occurred in the absence of international investment, rendering the task of awarding carbon credits for avoided deforestation (or new growth) highly uncertain. An inability to measure the carbon benefits of land-based mitigation with the precision of technological mitigation has impeded progress on the issue of forest protection for more than a decade.

Perhaps most important, forest ecosystems are simply too valuable to both regional and global ecological health to be subjected to the vagaries of a market mechanism for protection. Absent underlying regulation mandating investment in the protection and regeneration of global forests, unanticipated market forces can accelerate the destruction of forests, releasing in the process carbon stocks accumulated over centuries and eliminating a wide range of ecological services unrelated to carbon management. For these reasons, long-running attempts to develop a system of carbon credits fully fungible between technological and land-based mitigation programs are misguided. A more sound approach would require that nations participating in a global management framework commit not only to emission reductions from industrial processes but also to minimum levels of forest protection and regeneration in each commitment period.

The establishment of forest-management commitments by ratifying nations would sidestep many of the most significant obstacles to incorporating forest protection into the existing climate-management framework. Administered outside of carbon markets, forest-management programs would avoid altogether the thorny issue of "additionality," the extent to which investment in forest protection or regeneration yields a reduction in carbon beyond what would have occurred in the absence of such investment. Likewise, if maintained apart from carbon markets, forest-management contracts could vary in duration without concern for the future value of such contracts. In general, the precision with which the carbon benefits of forest protection and regeneration are measured would assume far less significance because such projects would be valued not only for sequestration but also for services nonreducible to a carbon-based metric, such as regional heat management.

The administrative structure of forest-management commitments could assume many forms, but the backbone of such a framework should be international investment in forest protection and regeneration in low- to mid-latitude regions where the climatic benefits of forest land covers are greatest. A range of criteria could be used to determine national commitment levels, with historical and present levels of emissions of greenhouse gases providing the most well-established basis for designating program obligations. Building on the Clean Development Mechanism of the Kyoto Protocol, the suitability of forest protection, reforestation, and agroforestry projects would need to be subject to an independent review process to ensure their compliance with performance goals related not only to climate management but also to a range of social and cultural concerns, such as the interests of indigenous populations, and other environmental considerations, such as water resources and wildlife management. Of course, total global investment in forest management would be necessarily governed by the willingness of nations with forested lands at low to mid-latitudes to participate in such a program.

To be most protective of human health, a percentage of forest-management commitments should be targeted toward urbanized regions. Enabling participant nations to direct these investments toward their own cities and to focus on a wider array of vegetative strategies than tree preservation and planting alone would provide a much stronger linkage between the global policy framework and local-scale planning than presently exists. Such projects would not need to be limited to tropical zones because vegetative enhancement in urban environments poses little risk of augmenting local temperatures, particularly in the warm season. Because urbanized regions account for only a small percentage of the global land surface, the proportion of investments directed to cities necessarily would be small but could yield substantial local benefits, likely generating greater carbon benefits per unit of area than nonurban projects.

Requiring the most highly emitting nations to invest in forest-management strategies abroad and in their own cities makes good sense not only as a means of enhancing regional climate resilience but also as a basis to secure emissions-reductions commitments from developing nations. With virtually all of the remaining tropical-forest area found within developing nations, the establishment of forest-management commitments would create a new and significant flow of investment from North to South, raising the potential to secure modest emissions-reduction commitments in exchange for sustained investments in forest projects. Such strategies should also be of interest to highly industrialized nations for which the cost of forest management is likely to be less than that of ever-deepening cuts to industrial emissions in future commitment periods.

4.   *Prioritize adaptive mitigation over nonadaptive mitigation.*

Recognition of the growing vulnerability of urban populations to extreme heat requires that the climate-management activities of municipal governments be broadened well beyond emissions reductions alone. One means of doing so is to prioritize investments in mitigation strategies that yield concurrent adaptive benefits over mitigation strategies that do not. At present, nonadaptive mitigation rules the day, with the vast majority of mitigation funds being directed to energy projects that produce no secondary benefits for local populations in the form of heat management, enhanced flood protection, or agricultural resilience. For example, mitigation strategies involving the substitution of a lower carbon-intensive fuel, such as natural gas, for a higher carbon-intensive fuel, such as coal, are an effective means of lowering CO_2 emissions, yet provide few other benefits related to climate management. A restructuring of the global policy framework to prioritize adaptive mitigation over nonadaptive mitigation, particularly in urban environments, would better integrate local and global policy objectives.

The international policy framework could be amended in several respects to better facilitate adaptive mitigation. One needed change is the inclusion of urbanized regions in national inventories of land-cover change. Under the Kyoto Protocol, national governments must track changes in net forest cover over time, information that is used to determine whether such changes are leading to an increase or reduction in forest-related emissions each year. Because the density of tree cover and other vegetation in cities typically does not meet the stated criteria for forest cover, urban environments are generally excluded from such inventories, unless national governments elect to account for changes in nonforest vegetative cover. A first step toward promoting adaptive mitigation in urban environments would be to require an inventory of land-cover changes in large cities at regular intervals, to assess how land-based mitigation strategies are influencing the distribution and extent of vegetative cover. Once in place, local vegetative strategies designed for heat management can be counted toward national emissions commitments under the Kyoto Protocol as well, providing a direct link between local and global policies.

A second policy change needed to facilitate adaptive mitigation is the development of carbon equivalencies for albedo-enhancement strategies. Of all the shortcomings of the global policy framework considered in this book, none is more significant than its failure to account for the nongaseous drivers of climate change. One such driver is the enhanced emittance of thermal radiation brought about through the urban-development process. Because both the emission of greenhouse gases and the release of thermal radiation contribute to rising temperatures, both should be the focus of mitigation activities under the global management framework. Yet, as a product of the Framework Convention definition of climate change, albedo enhancement has not been pursued as a mitigation strategy. A broadening of the formal definition of climate change to account for all physical drivers of warming, as recommended previously, would open the door to radiative mitigation strategies.

Accounting for the benefits of albedo enhancement under the Kyoto Protocol would require the development of carbon equivalencies for thermal radiation. Commonly referred to as "global-warming potential," such carbon equivalencies are used to determine how reductions in greenhouse gases other than carbon dioxide, such as methane or nitrous oxide, can be counted toward national emissions inventories, which employ carbon dioxide as a standard unit of measure. It is known, for example, that 1 ton of methane emissions is associated with the same magnitude of warming during the period of a century as 25 tons of carbon dioxide emissions, enabling mitigation strategies focused on methane reduction to be accounted for in carbon-equivalent inventories. A similar accounting scheme is possible for strategies focused on albedo enhancement, through which the warming potential of thermal radiation is mathematically equated with the warming potential of carbon dioxide [20]. Once established, land-based mitigation strategies that reduce the surface emittance of thermal radiation could be formally counted toward national carbon-reduction targets. The adoption of such equivalencies through an amendment to the Kyoto Protocol or a successor agreement would enable national governments to pursue urban albedo-enhancement strategies as part of their obligations under the international policy framework.

In concert, the inventorying of vegetative changes in cities and the adoption of a carbon equivalency for thermal radiation would enable national governments to account for the carbon-related benefits of adaptive mitigation strategies under the Kyoto Protocol. It is important to reiterate, however, that although tree planting and albedo enhancement in cities can counteract the global greenhouse effect, the more significant benefit of these strategies is the lowering of local temperatures through moderation of the urban heat-island effect. In this sense, although planting trees in cities carries with it a modest capacity for carbon sequestration, a much more immediate and powerful cooling effect is associated with an increase in local evapotranspiration and shading. Because the principal benefits of adaptive mitigation strategies tend to be noncarbon related, the global policy framework, with its emphasis on carbon management alone, discounts these approaches relative to nonadaptive mitigation. Revising the Kyoto Protocol to prioritize adaptive mitigation in cities would begin to address the growing need for adaptation in these environments through the existing policy framework.

One approach to promoting adaptive mitigation strategies under the Kyoto framework would be to require that a minimum percentage of a nation's carbon-management obligations be met through such strategies. Similar to the recommended requirement that a minimum percentage of forest-management programs be directed toward urban environments, such an approach would recognize the need for a growing share of climate-related investments to be directed toward adaptation in heavily populated areas. Alternatively, a differential accounting scheme could be developed for projects falling into specific categories of adaptive mitigation, enabling such projects to earn more emission credits per unit of controlled carbon than nonadaptive projects. In either case, the intent of such a policy change would be to redirect a proportion of new mitigation funds toward projects that yield local adaptive benefits in concert with emissions reductions. The extent to which adaptive mitigation strategies could be substituted for conventional mitigation necessarily would vary by region and climate zone.

Adopted as a package, these four recommended revisions to the established international framework for climate-change management would substantially elevate the significance of urban environments in global climate policy. Although these recommendations are designed to broaden the scope of climate-management activities beyond emissions control alone, there is good reason to suspect that an expanded focus on the land-surface drivers of climate change, both within and beyond cities, would have the effect of accelerating emissions-control efforts rather than slowing them. So great is the carbon-reduction potential of land-based strategies that even a pronounced shift from conventional to land-based mitigation would not be likely to undermine current carbon-control efforts. Yet, no such shift is needed. Rather, there is a pressing need for ongoing carbon-management activities to be augmented with a new set of climate-management tools if atmospheric carbon is to be drawn down to sustainable levels and urban environments are to remain habitable in a warming world. Broadening the established international framework to encompass land-based mitigation would advance both regional and global policy objectives.

It should be emphasized, however, that cities cannot rely on the global policy framework to govern regional-scale climate management. Fundamentally oriented toward the planetary-scale phenomenon of the global greenhouse effect, the Framework Convention and its associated protocols lack both the legislative mandate and regulatory scope to sufficiently alter the land-use practices of cities and regions. Indeed, even national governments may lack the direct regulatory authority needed to institute changes in the land-development and greenspace-planning practices of municipal governments. In many countries, it is municipal governments themselves that will need to institute policy changes and secure the needed resources to enhance regional climate resilience. A broadening of the global policy framework to encompass land-based mitigation facilitates regional climate management by establishing clear policy linkages between global and local management programs – by enabling, in effect, a leveraging of canopy for carbon. Yet, carbon control alone will not safeguard urban populations.

Recognition of the land-surface drivers of climate change alters the policy landscape in cities by empowering urban institutions to moderate the pace of regional warming without the need for broad intergovernmental coordination. In so doing, it effectively downscales the climate-change problem, reframing it in spatial and temporal terms that are far more meaningful to the daily lives of urban residents than the globalized climate narrative. More than any specific management technique, it is the contextualization of the climate-change problem at the scale of our own communities that carries with it the potential to mobilize support for climate initiatives commensurate with the immense challenge we confront. So, as with most significant societal challenges today, the problem of climate change entails both global and local dimensions, with action required on both fronts. As a result, managing climate change requires that we safeguard not only a distant global commons from rising levels of greenhouse gases; it also requires that we safeguard our own neighborhoods from rising levels of ambient heat. It requires, in short, that we confront climate change in the places we live.

Postscript

It has been my purpose in this book to describe the ways in which cities are altering their own climates and to consider what actions might contribute to a slowing of the rapid warming already underway in urban environments. The need for cities to actively manage the growing threat of extreme heat within their own jurisdictions has become ever more apparent during the period of the book's writing – a period in which cities around the world have experienced some of the most extreme climate-related events ever visited on human settlement. Yet, coupled with urban heat waves, droughts, flooding events, and tornadoes of unprecedented intensity in the past few years has been an almost complete breakdown of the international policy framework developed to manage the problem of climate change. In the wake of the pronounced failure of the Copenhagen Climate Conference to establish a successor agreement to the Kyoto Protocol, a growing list of developed nations have announced their intent to withdraw from the global policy framework altogether by the close of the first commitment period in 2012. These recent events in concert – a growing incidence of extreme weather in urban environments and a widening climate-policy vacuum – require cities to move more aggressively to protect their populations from climate-related threats in the present period.

It is my hope that this book can help inform the development of urban- and metropolitan-scale strategies to manage climate change more directly in the places most vulnerable to the effects of extreme heat. To this end, the book has been directed toward a few key purposes. The first of these has been to clearly differentiate global- from regional-scale climate-change processes and to do so without questioning the significance of global-scale climate change to cities. To those who may find in this book any ambivalence on my part as to the profound significance of the global greenhouse effect to the fate of cities or the planet as a whole, no such ambivalence is intended. As I describe in the book's first chapter, the mechanism of the global greenhouse effect has been widely validated and is generally accepted in the science community with the certainty of physical law. As asserted by a former head of the U.S. National Oceanic and Atmospheric Administration, only Newton's laws of motion may enjoy a wider scientific consensus than a human-enhanced greenhouse effect. And, like the laws of motion, an enhanced greenhouse effect is widely observable in the real world. More than three decades of data make plain a steady and unprecedented rise in observed temperatures measured at the global scale, along with more extreme-weather events and rising oceans along every continent – changes for which we must prepare regardless of their origins.

Yet, global-scale climate change is not the whole story. To observe the independent workings of well-established regional climate mechanisms is not to challenge the veracity of global-scale climate change; it is to assert only that the climate-change problem is more complex than the workings of the global greenhouse effect alone. It is also to assert that we need not rely on global institutions alone as we endeavor to protect our own communities from rising temperatures. For it is in this more complex understanding of the climate-change problem that novel management approaches are to be found, approaches that facilitate action across multiple geographic scales and through a broader range of governing institutions.

A second purpose of this book has been to argue that the regional-scale climate-change processes underway in urban environments are not unique to cities. Sometimes characterized in the climate-change literature as a minor climatic anomaly affecting only a small percentage of the global land surface, the urban heat-island effect is more accurately characterized as one end of a continuum across which land use influences climate. Land-use change is directly and profoundly influencing climate across large swaths of the global land surface through its effects on albedo and the surface energy balance, as well as on atmospheric carbon. Virtually all of the physical strategies available to counteract warming in cities are effective in mitigating climate change across much larger regions undergoing rapid land-use change. An understanding of how land use is driving climate in both urban and rural environments is essential to broadening the array of management tools available to moderate the impact of these changes on regional populations and ecosystems.

A third purpose of this book has been to argue that the present global policy framework for climate management has become an impediment to addressing the growing threat of extreme heat in cities. Although not the intent of its architects, the global policy framework impedes more effective action in cities by failing to address the principal drivers of climate change at the urban scale. Having characterized climate change in global terms only, the Framework Convention and its associated protocols have formalized a climate-change narrative that overlooks the land-based drivers of warming that, although regional in their origin and effects, are no less powerful drivers of climate change in cities than the global greenhouse effect. Rectifying this problem requires not that emissions-control efforts be scaled back in cities but rather that heat-management strategies be made far more robust – an outcome that could be achieved, in part, through revisions to the global policy framework.

Above all, this book has been concerned with the resilience of cities. If the development of the modern, post-industrial city was the defining legacy of the past century, it is the endurance of these places that will preoccupy us in the present one. Conceived in an era of limitless energy and a stable environment, the modern city is unlikely to enjoy either of these conditions in the coming decades. Therefore, the physical structure and management of cities must change if they are to persist in any sustainable form. As resilience to climate extremes becomes a chief design imperative in urban environments, cities will need to make extensive changes to the surface composition and spatial organization of their landscapes.

With this in mind, I have sought in this book to describe some of the physical changes needed in cities to address the growing vulnerability of urban populations to rapidly rising temperatures. These changes may be summarized as three general classes of urban climate-change management. The first set of strategies aggressively modifies the physical characteristics of cities to reflect or dissipate heat energy through enhanced surface reflection and vegetative cover. Such techniques involve an extensive expansion of urban tree canopies in climates supportive of tree growth, the direct incorporation of vegetative cover into building envelopes, and the widespread use of reflective materials in building and road resurfacing. In concert, these sunscreening strategies provide the most direct, immediate, and effective means of moderating the pace of warming in urban environments over time.

A second thrust of urban climate management addresses the regional drivers of urban-scale warming through restoring the ecological integrity of metropolitan hinterlands. In naturally forested regions, a cessation of deforestation in proximity to cities and an extensive reestablishment of forest cover can moderate the regional pace of warming trends. In arid regions, such strategies will require the protection and cultivation of the most moisture-efficient species of vegetation that can provide regional climatic benefits. In both instances, such greenbelting strategies will require a shift in metropolitan development from the rural periphery to the urban center, a shift further necessitated by rising energy costs.

A final thrust of urban climate management emphasizes energy conservation and efficiency to minimize the production of waste heat in cities. Enhanced energy efficiency in cities can slow the pace of warming by limiting the magnitude of fossil-fuel combustion for transportation, industry, and building climate control. By making use of conventional urban planning techniques ranging from enhanced building energy efficiency to the expansion of rail transit and pedestrian facilities, carbon-cooling strategies provide a direct linkage between regional- and global-scale climate management through limiting both the thermal and greenhouse emissions associated with fossil energy.

Rooted in the centuries of city-making that preceded the industrial era, these climate-responsive design techniques neither rely on the next wave of technological innovation nor require massive investments in decarbonized energy. They are techniques grounded in the same adaptive responses of ecological systems, and they are amenable to implementation across a range of scales, from that of the neighborhood to the metropolitan region. And, importantly, they are techniques that can be incorporated into the built environment in relatively little time. What such changes do require, however, is recognition on the part of urban governments of the seriousness of the challenge with which we are already confronted. To this end, the fundamental challenge posed by climate change in cities is revealed not as one of technological capacity but rather of governmental sufficiency: To what extent will cities choose to manage their own climate fate?

References

PROLOGUE

[1] "Sunshine hope for holiday weekend," *BBC News*, May 4, 2003, http://news.bbc .co.uk/go/pr/fr/-/2/hi/uk_news/2997435.stm.

[2] "Britain enjoys mini heat wave," *BBC News*, July 14, 2003, http://newsvote.bbc .co.uk/mpapps/pagetools/print/news.bbc.co.uk/2/hi/uk_news/3065199.stm.

[3] "France accused of complacency as 50 die in heatwave," *The Independent*, August 12, 2003.

[4] "Portugal declares disaster as fires kill nine: Heat wave continues," *Reuters*, August 5, 2003.

[5] "Crops at risk in France and Italy as drought, heat wave drag on," *Agence France Presse*, July 15, 2003.

[6] "Drought hit farmers plead for aid," *BBC News*, July 21, 2003, http://newsvote .bbc.co.uk/mpapps/pagetools/print/news.bbc.co.uk/2/hi/europe/3083625.stm.

[7] "Weather chaos across Europe: Breakfast's main story this morning was the chaos being caused by the weather across much of Europe," *BBC News*, July 31, 2003, http://news.bbc.co.uk/go/pr/fr/-/2/hi/programmes/breakfast/ 3112189.stm.

[8] "Buckling in the heat," *BBC News*, August 5, 2003, http://news.bbc.co.uk/2/hi/ uk_news/3126441.stm.

[9] "Helsinki tram schedules altered to give drivers more breaks during heat wave," *Nordic Business Report*, August 1, 2003.

[10] "Nuclear power plant has to be cooled with water spray," *The Herald* (Glasgow), August 5, 2003.

[11] *Impacts of Summer 2003 Heat Wave in Europe*, United Nations Environment Programme, March 2004, http://www.grid.unep.ch/product/publication/ download/ew_heat_wave.en.pdf.

[12] McCarthy, M. "The first decade: Man's race against time to halt climate change," *The Independent*, December 10, 2009.

[13] "It's getting hot in here," *BBC News*, August 2, 2004, http://newsvote.bbc.co. uk/mpapps/pagetools/print/news.bbc.co.uk/2/hi/uk_news/magazine/3527574. stm.

[14] Litchfield, J. "Holocaust of the elderly: Death toll in French heatwave rises to 10,000," *The Independent*, August 22, 2003.

[15] P. Pirard, S. Vandentorren, M. Pascal, et al., Summary of the mortality impact assessment of the 2003 heat wave in France, *Eurosurveillance*, 10, 153–5 (2005).

[16] F. Simon, G. Lopez-Abente, E. Ballester, et al., Mortality in Spain during the heat waves of summer 2003, *Eurosurveillance*, 10, 156–60 (2005).

175

[17] P. Michelozzi, F. de'Donato, L. Bisanti, et al., The impact of the summer 2003 heat waves on mortality in four Italian cities, *Eurosurveillance*, 10, 161–4 (2005).

[18] "French heatwave bodies kept in temporary storage," *The Independent*, August 25, 2003.

[19] "News roundup: Europe: Portugal counts cost of heatwave," *The Guardian*, August 26, 2003.

[20] Summer of forest fires and drought costs Europe's farmers," *Belfast News Letter*, October 18, 2003.

[21] "Prices rocket as veg supplies dry up in heatwave," *Western Daily Press*, September 12, 2003.

[22] "Chickens die in heat," *BBC News*, August 15, 2003, http://news.bbc.co.uk/go/pr/fr/-/2/hi/uk_news/england/lincolnshire/3153135.stm.

[23] J. Robine, S. Cheung, S. Le Roy, et al., Report on excess mortality in Europe during summer 2003, *EU Community Action Programme for Public Health* (Grant Agreement 2005114), 2007.

[24] A. Matzarakis, M. De Rocco, & G. Najjar, Thermal bioclimate in Strasbourg – the 2003 heat wave, *Theoretical and Applied Climatology*, 98, 209–20 (2009).

[25] Stott, P., Stone, D., & Allen, M. Human contribution to the European heatwave of 2003, *Nature* 432, 610–14 (2004).

[26] Jha, Alok. "Boiled alive," *The Guardian*, July 26, 2006.

CHAPTER 1

[1] Growing blanket of carbon dioxide raises Earth's temperature, *Popular Mechanics*, August, 119 (1953).

[2] G. Christianson, *Greenhouse: The 200 Year Story of Global Warming* (Walker and Co., 1999).

[3] S. Arrhenius, On the influence of carbonic acid in the air upon the temperature of the ground, *Philosophical Magazine and Journal of Science*, 41, 237–76 (1896).

[4] D. Archer, M. Eby, V. Brovkin, et al., Atmospheric lifetime of fossil-fuel carbon dioxide, *Annual Reviews of Earth and Planetary Sciences*, 37, 117–34 (2009).

[5] E. Kolbert, *Field Notes from a Catastrophe* (Bloomsbury, 2006).

[6] D. Lüthi, M. Le Floch, B. Bereiter, et al., High-resolution carbon dioxide concentration record 650,000–800,000 years before present, *Nature*, 453, 379–82 (2008).

[7] IPCC, *Climate Change 2001:The Scientific Basis. Contribution of Working Group I to the Third Assessment Report of the Intergovernmental Panel on Climate Change* (Houghton, J., Y. Ding, D. Griggs, M. Noguer, P. van der Linden, X. Dai, K. Maskell, & C. Johnson, eds.) (Cambridge University Press, 2001).

[8] *First session on the greenhouse effect and global climate change*, Committee on Energy and Natural Resources, United States Senate (S. HRG. 100-461 Pt 2), 1988.

[9] F. Wentz & M. Schabel, Effects of orbital decay on satellite-derived temperature trends, *Nature*, 384, 661–4 (1998).

[10] I. Allison, N. Bindoff, R. Bindschadler, et al., *Copenhagen Diagnosis: Updating the World on the Latest Climate Science* (University of New South Wales Climate Change Center, 2009).

[11] B. Lin, B. Wielicki, L. Chambers, et al., The iris hypothesis: A negative or positive cloud feedback?, *Journal of Climate*, 15, 3–7 (2002).

[12] K. Briffa, F. Schweingruber, P. Jones, et al., A reduced sensitivity of recent tree-growth to temperature at high northern latitudes, *Nature*, 391, 678–82 (1998).

[13] K. Briffa, T. Osborn, & F. Schweingruber, Large scale temperature inferences from tree rings: A review, *Global and Planetary Change*, 40, 11–26 (2004).

[14] IPCC, *Climate Change 2007: The Physical Science Basis. Contribution of Working Group I to the Fourth Assessment Report of the Intergovernmental Panel on Climate Change* (Solomon, S., D. Qin, M. Manning, Z. Chen, M. Marquis, K. B. Averyt, M. Tignor, & H. L. Miller, eds.) (Cambridge University Press, 2007).

[15] M. Blum & H. Roberts, Drowning of the Mississippi Delta due to insufficient sediment supply and global sea-level rise, *Nature Geoscience*, 2, 488–91 (2009).

[16] C. Hoyos, P. Adjudelo, P. Webster, et al., Deconvolution of the factors contributing to the increase in global hurricane intensity, *Science*, 312, 94–7 (2006).

[17] J. Elsner, J. Kossin, & T. Jagger, The increasing intensity of the strongest tropical cyclones, *Nature* 455, 92–5 (2008).

[18] M. Mann & L. Kump, *Dire Predictions: Understanding Global Warming* (Dorling Kindersley, 2008).

[19] M. Benton & R. Twitchett, How to kill (almost) all life: The end-Permian extinction event, *Trends in Ecology and Evolution*, 18, 358–65 (2003).

[20] J. Kiehl & C. Shields, Climate simulation of the latest Permian: Implications for mass extinction, *Geology*, 33, 757–60 (2005).

[21] C. Le Quere, M. Raupach, J. Canadell, et al., Trends in the sources and sinks of carbon dioxide, *Nature Geoscience*, 2, 831–6 (2009).

CHAPTER 2

[1] C. Marshall, R. Pielke, & L. Steyaert, Crop freezes and land use change in Florida, *Nature*, 426, 29–30 (2003).

[2] Sharp, D. "'Surprise freeze' burns farmers Fla. crop damage reaches $200M," *USA Today*, January 23, 1997.

[3] K. Miller, Response of Florida citrus growers to the freezes of the 1980s, *Climate Research*, 1, 133–44 (1991).

[4] C. Marshall, R. Pielke, L. Steyaert, et al., The impact of anthropogenic land-cover change on the Florida peninsula sea breezes and warm season sensible weather, *Monthly Weather Review*, 132, 28–52 (2004).

[5] "Tropical deforestation affects rainfall in the U.S. and around the globe," NASA, September 13, 2005, http://www.nasa.gov/centers/goddard/news/topstory/2005/deforest_rainfall.html.

[6] M. Williams, *Deforesting the Earth: From Prehistory to Global Crisis* (University of Chicago Press, 2003)

[7] P. Barreto, C. Souza, R. Nogueron, et al., "Human pressures on the Brazilian Amazon forest," World Resources Institute (2006), http://www.globalforestwatch.org/common/pdf/Human_Pressure_Final_English.pdf.

[8] R. Pielke, J. Adegoke, A. Beltran-Przekurat, et al., An overview of regional land-use and land-cover impacts on rainfall, *Tellus*, 59B, 587–601 (2007).

[9] P. Snyder, C. Delire, & J. Foley, Evaluating the influence of different vegetation biomes on the global climate, *Climate Dynamics*, 23, 279–302 (2004).

[10] M. Costa & J. Foley, Combined effects of deforestation and doubled atmospheric CO_2 concentrations on the climate of Amazonia, *Journal of Climate*, 13, 18–34 (2000).

[11] J. Feddema, K. Oleson, G. Bonan, et al., The importance of land-cover change in simulating future climates, *Science*, 310, 1674–8 (2005).

[12] R. Pielke, Land use and climate change, *Science*, 310, 1625–6 (2005).

[13] D. Olsen, Dividing Australia: The story of the rabbit-proof fence, *Things Magazine*, 14 (2001), http://www.thingsmagazine.net/text/t14/rabbits.htm.

[14] A. Pitman, G. Narisma, R. Pielke, et al., Impact of land cover change on the climate of southwest Western Australia, *Journal of Geophysical Research*, 109, D18109 (2004).

[15] D. Ray, U. Nair, R. Welch, et al., Effects of land use in Southwest Australia: Observations of cumulus cloudiness and energy fluxes, *Journal of Geophysical Research*, 108, 4414 (2003).

[16] G. Bonan, Forests and climate change: Forcings, feedbacks, and the climate benefits of forests, *Science*, 320, 1444–9 (2008).

[17] J. Juang, G. Katul, M. Siqueira, et al., Separating the effects of albedo from eco-physiological changes on surface temperatures along a successional chronose-quence in the southeastern United States, *Geophysical Research Letters*, 34, L21408 (2007).

[18] S. Roy, G. Hurtt, C. Weaver, et al., Impact of historical land cover change on the July climate of the United States, *Journal of Geophysical Research*, 108, 4793 (2003).

[19] R. Jackson, E. Jobbagy, R. Avissar, et al. Trading water for carbon with biological carbon sequestration, *Science*, 310, 1944–7 (2005).

[20] E. Kalnay & M. Cai, Impacts of urbanization and land use change on climate, *Nature*, 423, 528–31 (1997).

[21] E. Kalnay, M. Cai, H. Li, et al. Estimation of the impact of land-surface forcings on temperature trends in eastern United States, *Journal of Geophysical Research*, 111, D06106 (2006).

CHAPTER 3

[1] *Summary of Vital Statistics 2003: The City of New York* (New York City Department of Health and Mental Hygiene, 2004).

[2] D. Gaffen & R. Ross, Increased summertime heat stress in the U.S., *Nature*, 396, 529–30 (1998).

[3] B. Stone, J. Hess, & H. Frumkin, Urban form and extreme heat events. *Environmental Health Perspectives*, 118, 1425–8 (2010).

[4] P. Bosselmann, E. Arens, & K. Dunker, Urban form and climate: Case Study, Toronto, *Journal of the American Planning Association*, 61, 226–39 (1995).

[5] L. Howard, *The Climate of London* (W. Phillips & G. Yard, 1818).

[6] H. Landsberg, *The Urban Climate* (Academic Press, 1981).

[7] T. Oke, *Boundary Layer Climates* (Routledge, 1987).

[8] *The Growing Urbanization of the World*, Columbia Earth Institute (2005), http://www.earthinstitute.columbia.edu/news/2005/story03-07-05.html.

[9] *Extreme Heat: A Prevention Guide to Promote Your Personal Health and Safety*, U.S. Centers for Disease Control and Prevention, (2004), http://www.bt.cdc.gov/disasters/extremeheat/heat_guide.asp.

[10] C. Small, F. Pozzi, & C. Elvidge, Spatial analysis of global urban extent from DMSP-OLS night lights, *Remote Sensing of Environment*, 96, 277–91 (2005).

[11] R. Balling, R. Cerveny, & C. Idso, Does the urban CO2 dome of Phoenix, Arizona contribute to its heat island?, *Geophysical Research Letters*, 28, 4599–601 (2001).

[12] G. Ren, Z. Chu, Z. Chen, et al., Implications of temporal change in urban heat island intensity observed at Beijing and Wuhan stations, *Geophysical Research Letters*, 34, L05711 (2007).

[13] F. Fujibe, Detection of urban warming in recent temperature trends in Japan, *International Journal of Climatology*, 29, 1811–22 (2009).

[14] M. Kim, I. Kang, & C. Kwak, The estimation of urban warming amounts due to urbanization in Korea for the recent 40 years, *Journal of Korean Meteorological Society*, 35, 118–26 (1999).

[15] C. Yague, E. Zurita, & A. Martinez, Statistical analysis of the Madrid urban heat island, *Atmospheric Environment*, 25B, 327–32 (1991).

[16] R. Wilby, Past and projected trends in London's urban heat island, *Weather*, 58, 251–60 (2003).

[17] Y. Lim, M. Cai, E. Kalnay, et al., Observational evidence of sensitivity of surface climate changes to land types and urbanization, *Geophysical Research Letters*, 32, L22714 (2005).

[18] J. Tan, Y. Zheng, X. Tang, et al., The urban heat island and its impact on heat waves and human health in Shanghai, *International Journal of Biometeorology*, 54, 75–84 (2010).

[19] M. McCarthy, M. Best, & R. Betts, Climate change in cities due to global warming and urban effects, *Geophysical Research Letters*, 37, L09705 (2010).

CHAPTER 4

[1] United Nations Secretariat, *World Urbanization Prospects: The 2009 Revision* (Department of Economic and Social Affairs, 2010).

[2] A. Nelson, Leadership in a new era, *Journal of the American Planning Association*, 72, 393–407 (2006).

[3] American Forests, *Projected Environmental Benefits of Community Tree Planting: A Multi-Site Model Urban Forest Project in Atlanta* (2002).

[4] A. Rosenfeld, H. Akbari, J. Romm, et al., Cool communities: Strategies for heat island mitigation and smog reduction, *Energy & Buildings*, 28, 51–62 (1998).

[5] S. Solomon, G. Plattner, R. Knutti, et al., Irreversible climate change due to carbon dioxide emissions, *Proceedings of the National Academy of Sciences*, 106, 1704–9 (2009).

[6] H. Matthews & K. Caldeira, Stablizing climate requires near zero emissions, *Geophysical Research Letters*, 35, L04705 (2008).

[7] Y. Zhou & M. Shepherd, Atlanta's urban heat island under extreme heat conditions and potential mitigation strategies, *Natural Hazards*, 52, 639–68 (2010).

[8] M. McHale, E. McPherson, & I. Burke, The potential of urban tree plantings to be cost effective in carbon markets, *Urban Forestry & Urban Greening*, 6, 49–60 (2007).

[9] C. Rosenzweig, W. Solecki, L. Parshall, et al., *Mitigating New York City's Heat Island with Urban Forestry, Living Roofs, and Light Surfaces* (New York State Energy Research and Development Authority, 2006).

[10] C. Yu & W. Hien, Thermal benefits of city parks, *Energy and Buildings*, 38, 105–20 (2006).

[11] "Little green thumbs: Roof farms sprout on schools across New York City," *The Architect's Newspaper*, May 17, 2010.

[12] K. Liu & B. Bass, *Performance of green roof systems*, National Research Council Canada, (2005).

[13] K. Scholz-Barth, Green roofs: Stormwater management from the top down, Environmental Design and Construction (2001), http://www.edcmag.com/articles/green-roofs-stormwater-management-from-the-top-down.

[14] S. Peck & M. Kuhn, *Design guidelines for green roofs*, National Research Council Canada (2001).

[15] *Reducing Urban Heat Islands: Compendium of Strategies* (USEPA, 2008), http://www.epa.gov/heatisld/resources/compendium.htm.

[16] L. Stenning & S. Philipsen, *Seattle Green Factor: Improving Green Infrastructure Parcel by Parcel* (2008), http://greenfutures.washington.edu/powerpoints/Seattle_Green_Factor_Stenning_Philipsen.pdf.

[17] Witkin, J., "Developing a 'water-battery' for trees," *New York Times*, April 9, 2010.

[18] H. Taha, Urban climates and heat islands: Albedo, evapotranspiration, and anthropogenic heat, *Energy and Buildings*, 25, 99–103 (1997).

[19] H. Taha, S. Konopacki, & S. Gabersek, Impacts of large scale surface modifications on meteorological conditions and energy use: A 10-region modeling study, *Theoretical and Applied Climatology*, 62, 175–85 (1999).

[20] M. Hart & D. Sailor, Quantifying the influence of land-use and surface characteristics on spatial variability in the urban heat island, *Theoretical and Applied Climatology*, 95, 397–406 (2009).

[21] T. Oke, *Boundary Layer Climates* (Routledge, 1987).

[22] D. Sailor & L. Lu, A top-down methodology for developing diurnal and seasonal anthropogenic heating profiles for urban areas, *Atmospheric Environment*, 38, 2737–48 (2004).

[23] *Transportation Energy Data Book* (U.S. Department of Energy, 2010), http://cta.ornl.gov/data/tedb30/Edition30_Chapter02.pdf.

[24] B. Stone & M. Rodgers, Urban form and thermal efficiency: How the design of cities influences the urban heat island effect, *Journal of the American Planning Association*, 67, 186–98 (2001).

CHAPTER 5

[1] United Nations Framework Convention on Climate Change, May 9, 1992, S. Treaty Doc No. 102-38, 1771 U.N.T.S. 107.

[2] IPCC, *Climate Change 1995: The Science of Climate Change. Contribution of Working Group I to the Second Assessment Report of the Intergovernmental Panel on Climate Change* (Houghton, J., L. Filho, B. Callander, N. Harris, A. Kattenberg, & K. Maskell, eds.) (Cambridge University Press, 1996).

[3] IPCC, *Climate Change 2001: The Scientific Basis. Contribution of Working Group I to the Third Assessment Report of the Intergovernmental Panel on Climate Change* (Houghton, J., Y. Ding, D. Griggs, M. Noguer, P. van der Linden, X. Dai, K. Maskell, & C. Johnson, eds.) (Cambridge University Press, 2001).

[4] *Tracking Progress toward Kyoto and 2020 Targets in Europe*, European Environment Agency (Report No. 7, 2010), http://www.eea.europa.eu/publications/progress-towards-kyoto/.

[5] "Government confirms it will reject new Kyoto Protocol," *Reuters*, June 6, 2011, http://www.reuters.com/article/2011/06/09/us-climate-canada-idUSTRE75755O20110609.

[6] IPCC, *Climate Change 2001: Impacts, Adaptation, and Vulnerability. Contribution of Working Group II to the Third Assessment Report of the Intergovernmental Panel on Climate Change* (McCarthy, J., O. Canziani, N. Leary, D. Dokken, & K. White, eds.) (Cambridge University Press, 2001).

[7] J. Foley, M. Coe, M. Scheffer, et al., Regime shifts in the Sahara and Sahel: Interactions between ecological and climatic systems in Northern Africa, *Ecosystems*, 6, 524–39 (2003).

[8] L. Ornstein, I. Aleinov, & D. Rind, Irrigated afforestation of the Sahara and Australian Outback to end global warming, *Climatic Change*, 97, 409–37 (2009).

[9] W. Stibbe, J. Van der Koolj, J. Verweij, et al., *Response to Global Warming: Strategies of the Dutch Electricity Generating Board* (16th Congress of the World Energy Council, 1995).

[10] M. Oelbermann, R. Voroney, & A. Gordon, Carbon sequestration in tropical and temperate agroforestry systems: A review with examples from Costa Rica and southern Canada, *Agricultural Ecosystems and Environment*, 104, 359–77 (2004).

[11] *Paths to a Low Carbon Economy: Version 2 of the Global Greenhouse Gas Cost Abatement Curve* (McKinsey & Company, 2009), http://www.worldwildlife.org/climate/ WWFBinaryitem11334.pdf.

[12] "Distribution of registered project activities by scope," United Nations Clean Development Mechanism, http://cdm.unfccc.int/Statistics/Registration/ RegisteredProjByScopePieChart.html.

[13] M. Santilli, P. Moutinho, S. Schwartzman, et al., Tropical deforestation and the Kyoto Protocol, *Climatic Change*, 71, 267–76 (2005).

[14] IPCC, *Climate Change 2007: The Physical Science Basis. Contribution of Working Group I to the Fourth Assessment Report of the Intergovernmental Panel on Climate Change* (Solomon, S., D. Qin, M. Manning, Z. Chen, M. Marquis, K. B. Averyt, M. Tignor, & H. L. Miller, eds.) (Cambridge University Press, 2007).

[15] Africa 2000 Network Uganda, *2004 Annual Report*, http://www.a2n.org.ug/files/ 2004_Annual_Report_v0.7.pdf.

[16] L. Verchot, M. Noordwijk, S. Kandji, et al., Climate change: Linking adaptation and mitigation through agroforestry, *Mitigation and Adaptation Strategies for Global Change*, 12, 901–18 (2007).

[17] B. Chen, R. Mallick, & S. Bhowmick, A laboratory study on reduction of the heat island effect of asphalt pavements, *Journal of the Association of Asphalt Paving Technologists*, 78, 209–48 (2009).

[18] A. Max, "New energy uses for asphalt," *The Associated Press*, December 31, 2007.

[19] H. Akbari & S. Konpacki, Calculating the energy-saving potentials of heat-island reduction strategies, *Energy Policy*, 33, 721–56 (2005).

[20] G. Marland, R. Pielke, M. Apps, et al., The climatic impacts of land surface change and carbon management and the implications for climate-change mitigation policy, *Climate Policy*, 3, 149–57 (2003).

Index